Making Sure of Maths Book 4

T. F. Watson, M.A., F.E.I.S., and T. A. Quinn, M.A.

HOLMES McDOUGALL LIMITED

Printed & Published by Holmes McDougall Ltd, Edinburgh
Designed by Bob Crawford
Copyright © Holmes McDougall Limited, 1969

7157 0753-1

Contents

	Page		Page
Looking back	5	Mathematical sentences or equations	46
Fun with numbers	8	Indices	49
Time	11	Counting bases	53
Decimal fractions	14	Making sure 2	58
Length	18	Graphs	59
Area	21	Co-ordinates	63
Weight	25	More quadrilaterals	65
Proportion	27	Symmetry	69
Buying and selling	28	Sets	71
Percentages	30	Volume	75
Plans and Scales	33	Solids	79
Making sure 1	37	Making sure 3	83
Angles and tiles	38	Work cards	85
Circle patterns	42	Useful information	96
Drawing curves	44		

Looking back

A Rewrite the following sentences, using figures instead of words where possible. Remember to leave a space between millions and thousands and between thousands and hundreds.

1 There were ninety-eight thousand four hundred and seventy-five spectators at the match.

2 The attendance at the Cup Final at Hampden Park was one hundred and nineteen thousand seven hundred and fifty.

3 The population of Milchester is three hundred and twenty thousand five hundred and forty.

4 In the year nineteen hundred and sixty-six the population of Scotland was approximately five million one hundred and eighty thousand.

5 He was a very rich man and left one million eight hundred and seventy thousand four hundred and forty pounds.

B Write the number that is:

1 Fifteen more than seven thousand.

2 One hundred and nineteen more than ten thousand.

3 Six thousand and fifty more than ten thousand.

4 Fifty more than twelve thousand.

5 Ninety-eight more than one hundred thousand.

C The number 465 means 4 hundreds + 6 tens + 5 units.

1 What do these numbers mean:
206   438   2 537   14 012?

2 What is the value of the **2** in these numbers:
12   21   205   2 438   20 765?

3 What is the value of the **4** in these numbers:
4·6   40·2   8·4   408·9   7·24?

4 What is the value of the **5** in these numbers:
5·3   51·7   6·5   3·05   9·75?

5 Write the following in decimal form:
(a) 5 units + 3 tenths
(b) 2 tens + 5 units + 7 tenths
(c) 6 units + 5 tenths + 2 hundredths
(d) 4 units + 63 hundredths
(e) 7 units + 9 hundredths
(f) 3 units + 2 hundredths
(g) 29 hundredths
(h) 4 hundredths
(i) 8 tens + 5 hundredths
(j) one hundred + 6 hundredths.

6 Write these numbers to the nearest 10:
37   49   52   64   76
48   23   421   396   287

Write these to the nearest 100:
108   280   176   390   815
660   620   469   558   330

Write these to the nearest 1 000:
1 095   2 897   2 320   1 216
4 680   3 790   3 025   4 382

5

D True or False
Which of the following are true and which are false?

1. vi + v = xi
2. iv + vi = x
3. $6^2 + 4^2 = 10^2$
4. $7^2 + 6^2 = 8^2 + 5^2$
5. $(19 \times 6) + 6 = (5 \times 4) + 10^2$
6. $\frac{1}{4}$ is greater than $\frac{1}{8}$
7. $0.5 = \frac{5}{10} = \frac{1}{2}$
8. 0.9 is smaller than 0.09
9. 4 and 3 are factors of 24
10. 28 is a multiple of 7
11. $100 = 10^2$
12. 892 grammes is greater than 1 kilogramme.

E

1. A box holds $3\frac{1}{2}$ dozen pencils. How many pencils are there in 20 such boxes?

2. The population of Ashdale is 5 350. The population of Bigcroft is 4 672. If 265 people leave Bigcroft and go to live in Ashdale, what will be the population of each town then?

3. An aeroplane travels 720 kilometres in an hour. At this speed how far should it travel in 8 hours?

4. In a cricket match one team has scored 172 runs. When the other team has scored 89 runs, how many more runs must they score to win?

5. When 169 and another number are added the answer is 325. What is the number?

6. A farmer's field is this shape and size.

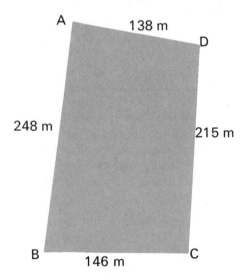

(a) What is the distance round it?
(b) Which is the shorter way to walk from corner A to corner C, via B or via D? How much shorter?
(c) Which is the shorter way from D to B, via C or via A? How much shorter?

7. 28 buses, each with 36 seats, carried the supporters of Elmpark Rovers to a match. All the buses were full except the last one which had seven empty seats. How many supporters were there?

8. When 1602 television valves were tested, 18 of them were found to be faulty. The rest were packed in boxes, each holding 24 valves. How many boxes were there?

9. A merchant had 600 boxes of chocolates. He sold $2\frac{1}{2}$ dozen to each of 18 shops.
(a) How many boxes did he sell?
(b) How many had he left?

10. When a number of children each took 8 caramels from a box holding 176 caramels, there were 16 left. How many children were there?

F Find the number that **n** stands for in these equations:

(1) n + 3 = 10
(2) n − 9 = 5
(3) n + 4 = 11
(4) n − 2 = 4
(5) n − 10 = 2
(6) n + 9 = 15
(7) n − 4 = 6
(8) n + 5 = 8
(9) n − 6 = 5
(10) n + 2 = 9
(11) n + 7 = 15
(12) n − 8 = 7

G Write the following, putting in the correct sign (>, <, =) in place of the question mark.

> means "is greater than."
< means "is less than."

1	6 × 9	?	9 × 6
2	(12 × 5) + 3	?	(13 × 5) − 2
3	(17 × 6) − 5	?	(18 × 5) + 5
4	(96 ÷ 8) + 10	?	(105 ÷ 7) + 6
5	(200 ÷ 8) + 5	?	(144 ÷ 6) + 10
6	(35 × 6) − 4	?	(39 × 5) + 4
7	(176 ÷ 4) + 6	?	(175 ÷ 5) + 9
8	£6·50	?	£6·05
9	£4·09	?	£4·10
10	£0·25	?	30p

H Find the **largest** whole number for **n** that will make these sentences true.

Then write the completed sentences. The first one is done for you.

1 5 × n < 12   (5 × 2 < 12)
2 6 × n < 34
3 7 × n < 42
4 9 × n < 60
5 8 × n < 89
6 2 × n < 21
7 5 × n < 40
8 6 × n < 23
9 4 × n < 43
10 7 × n < 47
11 3 × n < 17

I 1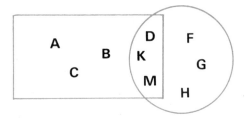

(a) Which letters are in the rectangle but not in the circle?
(b) Which letters are in the circle but not in the rectangle?
(c) Which letters are in both the circle and the rectangle?

2 Draw a rectangle and circle like the above.
(a) Write the numbers 1, 5, 9 in the rectangle but not in the circle.
(b) Write the numbers 2, 6, 8, in the circle but not in the rectangle.
(c) Write the numbers 3, 4, 7 in both the rectangle and the circle.

3 Write these sets in the proper mathematical way:
(a) the set of the first five whole numbers which divide evenly by 5.
(b) the set of tens between 29 and 71.
(c) a set of five fractions smaller than 0·5.
(d) a set of 4 metric measurements of length.

4 Here are the dates of birth of six people. Find their present ages in years and months:

M. Carr	9th November, 1926
A. White	14th March, 1932
T. Black	21st October, 1919
D. Jones	4th June, 1945
F. Smith	23rd May, 1939
H. Barr	15th July, 1941

7

Fun with numbers

1 Write down 1 number (any number) from each column of this table. Put them down as in an addition sum.

366	Take as an example these 5
147	numbers chosen at random
186	from each column.
179	Add up the units column
558	mentally: 6 + 7 + 6 + 9 + 8 = 36
———	Subtract 36 from 50. 50 − 36 = 14
1 436	The answer is **1 436**

69	345	186	872	756
366	642	582	278	558
168	246	87	575	657
762	147	285	377	954
663	543	483	179	855
564	48	384	674	459

Check your answer by adding in the usual way and you will see it is correct.

You can amaze your friends by this "lightning arithmetic".
A good idea is for you to add up the units figures as your friend is writing them down. By the time he has written the last one down you are ready to subtract from 50.

2

     (a)   (b)   (c)

Copy these squares and fill in the missing numbers.
Work from left to right and from top to bottom, taking each sign as it comes.

3 Copy these number crosswords and then fill in the missing numbers.

Across
a $\frac{1}{8}$ of 1 224
c $\frac{1}{4}$ of 104
e 431 − 379
g 236 + 127 + 144

Down
a £1·05 = ?p
b $6^2 − 2^2$
d 29 × 23
f 5^2

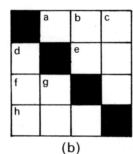
(a)

Across
a $\frac{1}{4}$ of 2 904
e 558 ÷ 6
f $3\frac{1}{2}$ dozen
h 9 × 8 × 12

Down
b 0·29 × 100
c 1 000 − 365
d $9^2 × 2^3$
g $6^2 − 10$

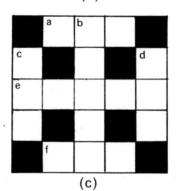
(b)

Across
a (173 × 4) − (6 × 9)
e 100 000 − 14 264
f 265 + 389 + 328

Down
b 50 000 − 14 232
c 73 × 8
d $10^2 + 9^2 + 7^2 + 6^2$

(c)

4 Draw this figure and then put the numbers below in the squares, so that every line of three numbers —up, across and diagonally— adds up to 60.

  12   14   18   20   22   26   28

Here's a hint.
What is the middle number?
Where should you put it?

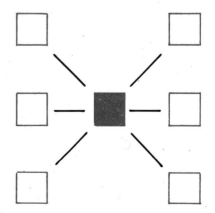

5 Ask your neighbour to write down three different numbers from 1 to 9, putting the smallest number first and the largest number last. Then ask him to:

    (a) Double the first number
    (b) Add 20
    (c) Multiply by 5
    (d) Add the second number
    (e) Multiply by 10
    (f) Add the third number
    (g) Subtract 1000

The answer should be the three numbers he wrote down.

6 It's fun to multiply by 9 if you do it this way.
Arrange 10 counters or any 10 things in a row.

To multiply 3 by 9 remove the 3rd counter. The answer is 27.

To multiply 5 by 9 remove the 5th counter. The answer is 45.
Try this with other numbers.

7 Here is a quick way to find the square of a 2-digit number which ends in 5.
    (a) $25^2 = 25 \times 25$. Multiply the 2 by the next highest number—3.
        $2 \times 3 = 6$ $5^2 = 5 \times 5 = 25$ So $25^2 = 625$
    (b) $35^2 = 35 \times 35$. Multiply the 3 by the next highest number—4.
        $3 \times 4 = 12$ $5^2 = 25$ So $35^2 = 1225$

Try these mentally: 15^2 45^2 55^2 65^2 75^2 85^2 95^2

8 A man has six pairs of brown socks and four pairs of black socks which he keeps in the same drawer. If he goes into the room in the dark, how many socks must he take from the drawer in order to make sure he has a pair to match?

9 I have two jugs. One holds 6 litres and the other 5 litres. Using these two jugs only, how can I get 2 litres into one of the jugs?

10 A woman hangs 4 wet shirts on a clothes line to dry. They dry in 20 minutes. How long does it take one shirt to dry?

Time

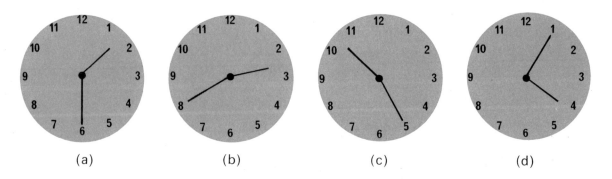

(a) (b) (c) (d)

The first two clocks above show morning (a.m.) times.
The second two clocks show afternoon and evening (p.m.) times.

Countries don't all have the same time. For example, when it is 12 noon here it is 7 a.m. in New York and 3 p.m. in Moscow. So we can say that New York time is 5 hours slow by our time, while Moscow time is 3 hours fast by our time.

The list below shows how many hours fast and slow times are in various places compared with our time. Write down the time it is in these places when our clocks show the times above.
If you don't know where all these places are, find out from your atlas.

Hours Fast			
Amsterdam	1	Singapore	$7\frac{1}{2}$
Bombay	$5\frac{1}{2}$	Sydney	10
Peking	8	Moscow	3

Hours Slow			
New York	5	Winnipeg	6
Rio de Janeiro	3	San Francisco	8
Buenos Aires	4	Hawaii	10

24-Hour Clock and Timetable

Travel timetables seldom use a.m. and p.m. Instead all times are 24-hour clock times. On this clock the hours from each midnight to the next are numbered from 0 to 24. Times from 1 p.m. onwards are shown by adding 12 hours.
Thus 2 p.m. is 14 hours, 3 p.m. is 15 hours.

To avoid mistakes, all times are shown by 4 figures. The first two figures give the hour and the last two figures the number of minutes past the hour.

4 p.m. is written as 16.00
3.15 p.m. is written as 15.15
6.5 p.m. is written as 18.05

9 a.m. is written as 09.00
7.45 a.m. is written as 07.45
8.5 a.m. is written as 08.05

A 1 Read these 24-hour clock times in the ordinary way:

06.25   05.40   09.35   16.00   18.26   19.05
08.28   07.05   15.00   20.25   20.45   22.25

2 Write these times as 24-hour clock times:

8 a.m.   9.45 a.m.   6.5 a.m.   7.2 a.m.   10.4 a.m.   9.20 a.m.
12 noon   1.35 p.m.   2.5 p.m.   8.50 p.m.   9 p.m.   11 p.m.

3 Write these 24-hour times as a.m. and p.m. times:

06.15   09.05   05.40   13.00   16.45   17.35
20.49   21.00   22.55   19.35   14.10   21.05

4 Find the number of hours and minutes from:

(a) 06.15 to 08.30   (b) 08.45 to 11.00   (c) 10.20 to 12.00
(d) 14.45 to 18.00   (e) 14.30 to 17.05   (f) 20.35 to 23.10

Check your answers by addition.

B Here is part of an Air Travel timetable.

Aberdeen — Edinburgh — London			
Aberdeen	depart	06.45	18.55
Edinburgh	arrive	07.25	19.35
,,	depart	07.50	20.15
London	arrive	09.10	21.35

1 How long is the forenoon flight from Aberdeen to Edinburgh?
2 How long does the plane stop at Edinburgh?
3 How long does it take to travel from Aberdeen to London?
4 How long from Edinburgh to London?
5 At what time (ordinary clock) does the afternoon plane leave Aberdeen?
6 How many minutes does it stop at Edinburgh?
7 At what time (ordinary clock) does this plane arrive in London?
8 How long does it take to travel from Aberdeen to London?
9 Copy the timetable but instead of the 24-hour clock times give ordinary clock times (a.m. and p.m.).

C   Here is a railway/steamer timetable from London to Ostend, in Belgium.

London—Dover—Ostend	Train 1	Train 2	Train 3
London .... depart	09.00	14.00	23.00
Dover .... arrive	10.22	15.22	00.22
„ .... depart	11.00	16.00	01.15
Ostend .... arrive	15.20	20.20	05.15

1 How long does it take Train 1 to travel from London to Dover?
2 Do Train 2 and Train 3 take the same time as Train 1 to travel to Dover?
3 If you go by Train 1 to Dover, how much time do you spend there?
4 Would you spend the same time in Dover, if you went by Train 3?
5 If you left London by the morning train, how long would it take you to reach Ostend?
6 At what time (ordinary time) would you arrive in Ostend?
7 If you travelled by bus to Dover and arrived at half-past 2 in the afternoon, how long would you have to wait for a steamer?
8 At what time (ordinary clock) should this steamer arrive in Ostend?

D   Here is an old railway timetable. Re-write it in the modern way, that is, giving the times as 24-hour clock times.

Edinburgh .... depart	8 a.m.	10 a.m.	12 noon
Berwick-upon-Tweed „	8.55	—	—
Newcastle .... „	10.0	11.55	2.4 p.m.
Darlington .... „	10.46	—	2.53 p.m.
London .... arrive	2.15 p.m.	4.0 p.m.	7.45 p.m.

1 Which train does the journey from Edinburgh to London in the shortest time?
2 Which is the slowest train?
3 Which is the only train stopping at Berwick-upon-Tweed?
4 How long does this train take between Berwick and London?
5 How long does the afternoon train from Darlington take to reach London?
6 If this train is 18 minutes late, when does it arrive in London?

Decimal fractions

A Addition and subtraction

We have already learned that when adding or subtracting decimals we must put the points under each other, making sure that the point in the answer is under the other points.

```
   5·8
  12·25
   4·72
 + 6·5
 ─────
  29·27
```

Now try these:

(a)
1 2·6 + 4·75 + 6·3
2 6·2 + 3·06 + 9·64
3 3·97 + 6·23 + 5·2
4 6·38 + 0·46 + 12·4

(b)
10·52 + 1·05 + 0·76
9·48 + 0·06 + 5·95
7·14 + 6·87 + 6·25
3·95 + 4·36 + 7·78

(c)
20·2 − 6·5
17·0 − 8·45
21·3 − 3·58
12·5 − 7·69

B

1 Add seven point nine to eight point five four.
2 After spending 0·2, 0·25 and 0·18 of my money, what fraction of my money have I left?
3 Tom cycled 18·25 kilometres. Jack cycled 15·8 kilometres. How much farther did Tom cycle?
4 From a piece of wallpaper 4 metres long, a piece 0·75 metres long was cut off. How much was left?
5 Three ropes tied together measured 30·75 metres. One rope was 9·6 metres long and a second rope was 8·9 metres. What was the length of the third rope?
6 A shop had 50 metres of tweed. One customer bought 8·75 metres and another customer bought 12·5 metres. How much was left?
7 An oil drum contained 62 litres of oil. I took out 15·4 litres one day and 9·65 litres the next day. How much oil was left in the drum?
8 What must be added to (6·42 + 3·76 + 5·95) to make 19·5?
9 A parcel contains 0·05 kg of chocolates, 0·16 kg of caramels and 1·35 kg of mixed biscuits. How much short of 2 kg does it weigh?
10 How many litres of milk are left if 2·5 litres and 1·65 litres are poured into a large jug and later 1·75 litres are poured out?
11 What must be added to 3·46 kg to make up $5\frac{1}{2}$ kg?
12 A boy wrote £0·50 instead of £0·05. By how many pence was his answer wrong?

C Multiplication by a whole number

We multiply in the usual way, making sure that the decimal point in the answer is placed directly below the point in the number being multiplied.
Note that 2·85 × 5 = 5 × 2·85

```
  2·85
   ×5
 ─────
 14·25
```

Multiply:
(a) 1·8 by 3, by 4 and by 5
(b) 3·45 by 5, 6 and 7
(c) 7·05 by 7, 8 and 9
(d) 6·25 × 7
(e) 6 × 4·62
(f) 9 × 17·75
(g) 4 × 9·25
(h) 8 × 5·46
(i) 7 × 8·95

D **Multiplying by 10, 20, 30 and so on**

In earlier books we learned that to multiply by 10, we move the figures one place to the left, and to multiply by 100 we move them two places to the left.

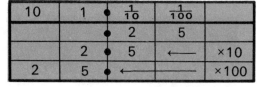

10	1	$\frac{1}{10}$	$\frac{1}{100}$	
		2	5	
	2	5		×10
2	5			×100

Multiply by 10:   (a) 0·34   (b) 0·5   (c) 2·4   (d) 3·25   (e) 4·05
Multiply by 100:   (f) 0·35   (g) 0·6   (h) 1·8   (i) 2·56   (j) 3·02

To multiply by 20, 30 and so on, multiply by 10 first, then by 2, 3 and so on

$$3·2 \times 20 = 32 \times 2 = 64$$

To multiply by 200, 300 and so on, multiply by 100 first, then by 2, 3 and so on

$$3·2 \times 300 = 320 \times 3 = 960$$

(a) Now try these:
(1) 2·3 × 20   (2) 5·3 × 40   (3) 0·6 × 60
(4) 1·7 × 30   (5) 0·2 × 50   (6) 3·9 × 40

(b) Now try these:
(1) 1·24 × 200   (2) 2·08 × 300
(3) 2·15 × 500   (4) 0·08 × 400
(5) 0·75 × 200   (6) 1·25 × 300

E **Extending the decimal number system**

In the table above we counted down to hundredths. We can go further and count up to thousands and down to thousandths.

TH	H	T	U	t	h	th
1 000	100	10	1	$\frac{1}{10}$	$\frac{1}{100}$	$\frac{1}{1\,000}$
			2	6	4	5

$2·645 = 2$ wholes $+ \frac{6}{10} + \frac{4}{100} + \frac{5}{1\,000}$ (or $\frac{645}{1\,000}$)

F **Multiplication by a decimal**

$0·9 \times 0·7 = \frac{9}{10} \times \frac{7}{10} = \frac{63}{100} = 0·63$

There is a total of 2 figures after the point in the numbers being multiplied, and there are 2 figures after the point in the answer.

```
   6·37    2 figures after the point
 × 2·5     1 figure after the point
  3185
  1274
 15·925    So 3 figures after the point
           in the answer.
```

Before multiplying the following, decide on an approximate answer to each. Make sure you have the right number of figures after the point in the answer:

(1) 4·8 × 0·4   (2) 5·09 × 0·2
(3) 16·7 × 0·4   (4) 0·95 × 0·5
(5) 0·05 × 2·3   (6) 12·6 × 2·3
(7) 10·04 × 1·6   (8) 1·05 × 6·4
(9) 8·04 × 2·5   (10) 0·94*l* × 0·5
(11) 0·85*l* × 0·3   (12) 1·25*l* × 0·5
(13) 0·78*l* × 0·3   (14) 3·25 m × 2·5
(15) 2·5 cm × 3·2   (16) 4·86 kg × 1·5
(17) 6·4 kg × 3·6   (18) 6·5 km × 2·4

1 What is the total length of 9 planks, each 3·5 metres long?

2 What is the total weight of 50 boxes, each weighing 0·75 kilogrammes?

3 I have £1·25. Dad has 15 times as much. How much money has he?

4 Find the cost of 24 metres of cloth at £1·23 per metre.

5 What would be the cost of 8·5 metres of cotton at £0·75 per metre?

G Division by a whole number

> Divide 3·15 metres into 5 equal parts
> $$\frac{0·63}{5)3·15}$$
> Each part is 0·63 metres.

We divide as in whole numbers. We put the decimal point in the answer directly above the point in the number being divided. This is important.

Now try these:

(1) 37·5 ÷ 5 (2) 8·28 ÷ 4 (3) 17·04 ÷ 3 (4) 82·4 ÷ 8
(5) 57·6 ÷ 9 (6) 5·85 ÷ 9 (7) 87·5 ÷ 5 (8) 0·96 ÷ 12
(9) 38·76 ÷ 12 (10) 19·84 ÷ 16 (11) 28·8 m ÷ 9 (12) £53·48 ÷ 7
(13) £17·50 ÷ 14 (14) 19·84 kg ÷ 8 (15) 77·76 kg ÷ 24

Division by 20, 30 and so on

(a) We learned in Book 3 that to divide by 10, we move the figures 1 place to the right, and to divide by 100 we move them 2 places to the right.

10	1	•	$\frac{1}{10}$	$\frac{1}{100}$	
2	5	•			
→	2	•	5		÷ 10
→	→	•	2	5	÷ 100

Divide by 10: 34   23·8   41·5   2·4   0·6   0·92
Divide by 100: 234 320   26·2   20·4 12·3 10·4

(b) To divide by 20, 30 and so on, first divide by 10, then by 2, 3, and so on.

Now try these:
(1) 363 ÷ 30 (2) 28·8 ÷ 40 (3) 14·4 ÷ 60 (4) 16·8 ÷ 20
(5) 175 ÷ 50 (6) 468 ÷ 30 (7) 17·6 ÷ 40 (8) 276 ÷ 20

H Dividing a decimal by a decimal

It is always easy to divide by a whole number, so when we have to divide by a decimal first change the decimal to a whole number.

(a) To make 0·6 a whole number we multiply by 10.   So 0·6 × 10 = 6
    To make 1·2 a whole number we multiply by 10.   So 1·2 × 10 = 12
    To make 0·25 a whole number we multiply by 100. So 0·25 × 100 = 25

By what must you multiply these numbers to make them whole numbers?

(1) 0·4   (2) 0·5   (3) 1·8   (4) 0·45   (5) 0·08   (6) 1·05

(b) $\frac{2}{5} = \frac{4}{10}$ (multiplying both numerator and denominator by 2)

$\frac{0·4}{0·2} = \frac{4}{2}$ (multiplying both numerator and denominator by 10)

You see that *multiplying both numerator and denominator by the same number does not change the value of the fraction.*

$$\frac{0.15}{0.3} = \frac{0.15 \times 10}{0.3 \times 10} = \frac{1.5}{3} = 0.5 \qquad \frac{8.1}{0.09} = \frac{8.1 \times 100}{0.09 \times 100} = \frac{810}{9} = 90$$

Now try these:

(1) $37.5 \div 0.5$ (2) $26.4 \div 0.3$ (3) $3.86 \div 0.2$ (4) $3.35 \div 0.5$

(5) $3.68 \div 0.8$ (6) $2.45 \div 0.7$ (7) $0.49 \div 0.07$ (8) $0.56 \div 0.08$

(9) $1.92 \div 0.12$ (10) $4.2 \div 0.35$ (11) $\dfrac{2.4}{0.04}$ (12) $\dfrac{21.6}{2.4}$

(13) $\dfrac{90}{1.5}$ (14) $\dfrac{0.036}{0.06}$ (15) $\dfrac{0.125}{0.05}$

1. How many lengths of 1·2 metres can be cut from a piece of wood 8·4 metres long?

2. How many skipping ropes, each 2·4 metres long, can be cut from a rope which measured 19·2 metres?

3. My step is 0·9 metres. How many steps do I take in walking 81·9 metres?

4. How many 0·6 kg bags of sugar can be filled from a sack containing 10·8 kg?

5. 96 metres of rope were cut into pieces 4·8 metres long. How many pieces were there?

6. The sum of £38·25 was shared equally among a number of men. Each man received £2·25. How many men were there?

7. How many times is £0·85 contained in £10·20?

8. How often can a jug which holds 0·6 of a litre be filled from a container which holds 22·2 litres?

9. How many 0·25 kg bars of chocolate would weigh 7·5 kg?

10. How many pieces of carpet 1·5 metres long can be cut from a length of 6 metres?

11. How many monthly payments of £2·40 are needed to pay a debt of £14·40?

12. How many suits, each requiring 5·8 metres of cloth, can be cut from a length of 52·2 metres?

13. A workman earns £5·56 a day. How many days did he work to earn £50·04?

14. How many parcels each weighing 0·75 kg can be made up from 3·75 kg of fruit?

MATHS 4–B

Length

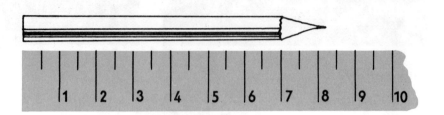

We have already learned that:

(a) the principal unit of length is the *metre*
(b) 1 metre (m) = 100 centimetres
(c) 1 centimetre (cm) = 10 millimetres (mm)
(d) 1 metre = 1 000 millimetres
(e) 1 000 metres = 1 kilometre (km)

(a) —————————— (b) ——————————
(c) —————————— (d) ——————————
(e) —————————— (f) ——————————

Measure these lines. Write your measurements like this:
(a) 5 cm 7 mm = 5·7 cm = 57 mm
Draw lines of   4·5 cm   5·8 cm   6·9 cm   75 mm   88 mm

Since centimetres are hundredths of a metre,
1 metre 45 centimetres may be written as 1·45 m
2 metres 40 centimetres as 2·40 m
2 metres 4 centimetres as 2·04 m

A
1 Write as shown above:
(a) 4 metres 25 centimetres   (b) 6 m 56 cm   (c) 3 m 40 cm   (d) 3m 4 cm

2 How many centimetres in:
(a) 2 metres   (b) 3 metres   (c) 1½ metres   (d) 3·25 m   (e) 2·75 m   (f) 4·36 m?

3 How many millimetres in:
(a) 2 centimetres   (b) 3 cm   (c) 1·5 cm   (d) 2·9 cm   (e) 4·7 cm   (f) 0·5 cm?

4 How many metres are equal to:
(a) 200 cm   (b) 500 cm   (c) 375 cm   (d) 580 cm   (e) 1 000 cm   (f) 2 500 cm?

5 How many centimetres are equal to:
(a) 1 000 mm   (b) 950 mm   (c) 480 mm   (d) 695 mm   (e) 375 mm   (f) 205 mm?

6. A man is 185 cm tall. What is his height in metres?
7. A girl is 1·52 m tall. What is her height in cm?
8. A motor car has these dimensions: Length 4·24 m, Width 1·70 m, Height 1·42 m. Write these measurements in cm and also in mm.
9. The dimensions of a French car are: Length 3 800 mm, Width 1 445 mm, Height 1 345 mm. Write these dimensions in m.
10. Find your height in
    (a) m   (b) cm   (c) mm
11. Measure other heights and write them in (a) m   (b) cm   (c) mm.
12. Use a metre stick or other metre measure to find to the nearest ½ m (0·5 m) the length and the breadth of your classroom.

B 1. Three pencils measuring 17·5 cm, 15·8 cm and 12·6 cm are placed end to end on a table. What is the total length?
2. After 3 lengths of 4 m, 6·55 m and 5·75 m have been cut from a roll of cloth there is a length of 12·10 m left. What was the length of the roll at first?

3. From a piece of string 5 metres long a length of 3·25 metres is cut off. How many centimetres of string are left?
4. Jim is 35 cm smaller than his father who is 1·85 m tall. What is Jim's height?
5.
   A field is triangular in shape.
   (a) What is the distance all round the field?
   (b) By how much is the longest side greater than the shortest side?
6. I had 3 pieces of string measuring 6·25 m, 5·98 m and 7·75 m. I joined them together, wasting 10 cm in tying the knots. What was the length of the knotted string?

7. What is the total length of 4 pieces of wire, each 56 cm long?

8. Some books are placed one on top of another. There are 9 books, each 4·5 cm thick and 8 books each 3·6 cm thick. How high is the pile of books?
9. A rope 25 m long was cut into 4 equal pieces. How long was each piece?
10. A boy's step is 54 cm. How many metres does he walk in 60 steps?
11. A man's step is 69 cm. How many steps does he take in walking 34·5 metres?
12.

    What is the length all round the carpet?
13. How many lengths of 75 cm can be cut from a rope 9 m long?
14. A boy was climbing up a pole 18 m high. When he had climbed 13·25 m he slipped back 2 m 50 cm. How far was he then from the top?

C Longer lengths such as the distances between two towns are measured in *kilometres* (km). *Kilo* means 1 000, so 1 km = 1 000 m.

```
          1 km = 1 000 m                          1 000 m = 1 km
So        3 km = (3 × 1 000) m          So  2 175 m = (2 175 ÷ 1 000) km
               = 3 000 m                                = 2·175 km
and 2·750 km = (2·750 × 1 000) m        and 3 600 m = (3 600 ÷ 1 000) km
               = 2 750 m                                = 3·600 km
```

1 It is easy to multiply decimal fractions by 1 000, so in one step change to metres:
2 km   2·635 km   3·890 km
1·500 km   6·176 km   4·305 km

2 In one step change the following to kilometres:
3 670 m   2 800 m   5 365 m
1 965 m   4 268 m   2 507 m

A distance may be expressed as 0·600 km or as 0·6 km. There is really no difference.

$$0·6 \text{ km} = \tfrac{6}{10} \text{ km} = \tfrac{6}{10} \text{ of } 1\,000 \text{ m} = 600 \text{ m}$$

Similarly $0·75 \text{ km} = \tfrac{75}{100} \text{ km} = \tfrac{75}{100}$ of 1 000 m = 750 m

3 A runner in an 8 kilometres race was forced to drop out when he was 462 metres from the finish. What distance had he run?

4 A bus makes 4 return journeys every day between two villages which are 12·250 km from each other. What distance does it cover each day?

5 Jack set out to walk to his uncle's house which was $7\tfrac{1}{2}$ km away. After he had walked 5 km 46 m, how far had he still to walk?

6 A man travelled 126·5 km by train, 7 km 250 m by bus, and walked 875 m. How far did he travel altogether?

D

Aberdeen

338	Carlisle							
188	150	Edinburgh						
228	151	71	Glasgow					
167	404	254	272	Inverness				
496	187	307	338	562	Leeds			
528	188	338	341	594	117	Liverpool		
130	220	71	98	185	378	409	Perth	
484	182	296	333	550	39	156	367	York
788	480	600	631	854	306	317	671	317	London

Distances are in kilometres. They are approximate distances by road.

The table shows the approximate distances between several towns. It is very easy to read.

Example: Find the distance from Leeds to York.

Where the Leeds **column** and the York **row** meet you will find the number 39. The distance is 39 kilometres.

Now find the distance from:

(1) Aberdeen to   (a) Glasgow   (b) London
(2) Edinburgh to   (a) Leeds   (b) York
(3) Liverpool to   (a) Carlisle   (b) Aberdeen
(4) How much further is it from Carlisle to London than to Inverness?
(5) How much further from Perth to London than from Aberdeen to York?

Area

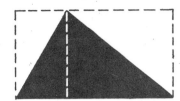 

In Book 3 we saw that the area of a surface, no matter its size and shape, is measured in square units. The area of small surfaces, such as a page of this book, is measured in square centimetres (for short cm^2). Larger areas, such as the surface of a table, may be measured in squares which have sides of 10 cm.

Draw a centimetre square and a 10 cm square.

How many square centimetres are there in a 10 cm square?

It would take a very large number of either of these squares to measure the area of the floor of your classroom or of the playground,
so we use a square whose sides are 1 metre long.

This we call a square metre (m^2).

To get an idea of the size of a square metre place 4 metre sticks on the floor to make a square. The area enclosed by the sticks is almost a square metre.

Estimate (guess) the number of square metres in your classroom floor.

Very large areas like the area of a town, a county or a country, are measured in square kilometres (km^2).

We have already seen that: **Area of a rectangle = Length × Breadth**

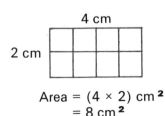 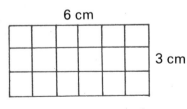 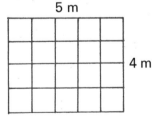

Area = (4 × 2) cm^2
     = 8 cm^2

Area = (6 × 3) cm^2
     = 18 cm^2

Area = (5 × 4) m^2
     = 20 m^2

Remember that the length and breadth must both be in centimetres, or both in metres.

A 1 Find the area of the following squares and rectangles:

	Length	Breadth		Length	Breadth
(a)	13 cm	11 cm	(f)	16 m	8 m
(b)	15 cm	9 cm	(g)	16 m	$8\frac{1}{2}$ m
(c)	12 cm	12 cm	(h)	14 m	14 m
(d)	16 cm	12 cm	(i)	20 m	$10\frac{1}{4}$ m
(e)	10 cm	$9\frac{1}{2}$ cm	(j)	16 m	$12\frac{1}{4}$ m

2 Find the area of a rectangular garden 70 metres long and 50 metres broad.

3 How many square tiles of side 10 cm are needed to cover an area 3 m by 2 m?

4 Find the area of a lady's scarf 60 cm square.

5 A lawn is 36 m long and 20·5 m broad. Find its area.

6 The living room of a house is 6 m by 4·5 m. A bedroom is 5·5 m by 4 m. Which room has the more floor space and by how many square metres?

7 Which has the greater area, and by how many square metres:
(a) a plot 18·5 m by 12 m, or (b) a plot 20 m by $11\frac{1}{2}$ m?

8 Rectangles 16 m by 1 m, and 8 m by 2 m, and a square 4 m by 4 m all have the same area. Find the lengths and breadths of four rectangles and one square all of which have an area of 36 square metres.

B We saw that the diagonals of a square and of a rectangle divided each shape into two equal triangles. Each triangle has half the area of the square or rectangle.

Look at these three triangles:

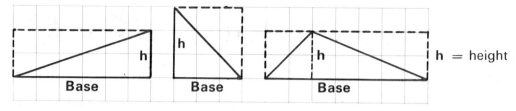

The area of each square or rectangle = base × height.
The area of each triangle = $\frac{1}{2}$ of the base × height.

1 Find the areas of these triangles.

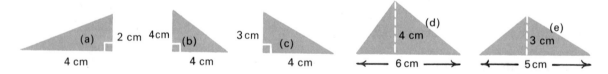

2 If each small square stands for 1 square cm, find the areas of these shapes. The dotted lines will help you.

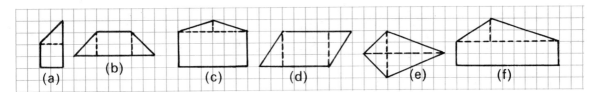

C Carpets and linoleum are generally sold by the square metre.
If we have a carpet, we often lay linoleum round it to form a border.

Here we see a drawing of the floor of a room which is 5 m by 4 m. A carpet, 4 m by 3 m, is laid on the floor. We want to find how many square metres of floor are not covered by the carpet. We can cover this area with linoleum.

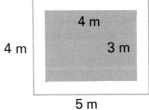

We can find the area we want by finding the area of the carpet and then subtracting this area from the area of the floor, like this:

> Area of floor = (5 × 4) m² = 20 m²
> Area of carpet = (4 × 3) m² = 12 m²
> Area not carpeted = 20 m² − 12 m² = 8 m²

So that we would need 8 square metres of linoleum.

Now find these areas. Draw a rough plan, if it will help you.

1 A carpet 5 metres long and 4 metres wide is laid on a floor 6 metres long and 5 metres wide. What area is not carpeted?

2 A rug 2 m long and 1·5 m wide is laid on the floor of a small bedroom which is 4 m long and 3·5 m wide.
 (a) How many square metres are covered by the rug?
 (b) To cover the rest with linoleum, how many square metres would you need?

3 A lawn, 24 metres by 20 metres, has a rectangular rose bed, which is 8 metres by 7 metres in the middle of it.
How many square metres of turf are there in the lawn?

4 A rectangular fish pond, 3 m by 2 m, has a cemented path all round it. The path is 2 m wide.
 (a) Draw a rough plan and put in the dimensions of the path and pond.
 (b) Find the area of the path.

5 A framed picture is 40 cm by 32 cm. The actual picture measures 36 cm by 24 cm. What is the area of the border?

6 Here are some rectangular shapes with the measurements shown. In each figure, find the area of the coloured border in square metres.

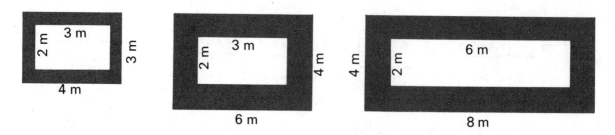

D We found that 6 × 3 = 18 from our multiplication square.
We also found that this means as well that:

$$18 \div 6 = 3 \text{ and } 18 \div 3 = 6$$

Here we have a rectangle 6 metres long and 3 metres wide. The area is (6 × 3) m² = 18 m².

In the same way, 18 ÷ 6 = 3, and so the width is 3 metres.
And 18 ÷ 3 = 6, and so the length is 6 metres.

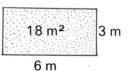

1 Look at this rectangle. Its area is 24 square centimetres.
   (a) How many square cm are there in the top row?
   (b) How many rows must there be to give an area of 24 square cm for the whole rectangle?

2 (a) If the area of a rectangle is 35 square cm and there are 7 square cm in the top row, how many rows of 7 square cm are there in the rectangle?
   (b) What is the breadth of the rectangle?

3 Copy this table and write down the missing measurement.

	Area	Length	Width		Area	Length	Width
(a)	36 cm²	9 cm	?	(g)	150 m²	15 m	?
(b)	54 cm²	?	6 cm	(h)	220 m²	20 m	?
(c)	96 cm²	12 cm	?	(i)	126 m²	?	9 m
(d)	63 cm²	?	7 cm	(j)	?	12 cm	$6\frac{1}{2}$ cm
(e)	84 cm²	12 cm	?	(k)	?	20 cm	8·5 cm
(f)	40 cm²	?	5 cm	(l)	112 m²	?	8 m

Weight

We have already learned that the weights usually used are the gramme (g) and the kilogramme (kg).

The gramme is a very small weight.

Kilo means 1 000, so 1 000 grammes = 1 kilogramme.

Since a gramme is a thousandth of a kilogramme,
(a) 1 kg 450 g may be written as 1·450 kg and 2 kg 5 g as 2·005 kg
(b) 2 165 g may be written as 2·165 kg and 1 250 g as 1·250 kg.

1 Write as kg:
2 000 g; 1 725 g; 3 475 g; 524 g.

2 Write as grammes:
3 kg; 1·346 kg; 2·006 kg; $2\tfrac{1}{2}$ kg.

3 What must be added to 572 g to make
(a) 1 kg   (b) $1\tfrac{1}{2}$ kg   (c) 2·100 kg?

4 I posted 2 parcels. One weighed 946 g and the other 758 g. What was the total weight in kg?

5 An empty box weighs 1·250 kg. When 18·460 kg of oranges are put into it what will it weigh?

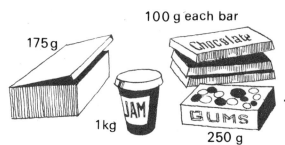

6 I sent my cousin a parcel made up of the articles above.
(a) What was the weight of the parcel?
(b) How much less than 3 kg was it?

7 A merchant placed 2 loads on his van. One weighed 98·550 kg and the other 174·685 kg. What weight must be added to make 300 kg?

8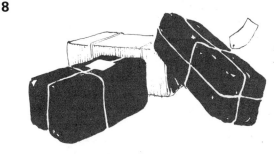

What is the total weight of the parcels if each weighs 2 kg 450 g?

9 If I weighed 10·850 kg more I would be the same weight as my mother who weighs 57·645 kg. What is my weight?

10 A box can hold $2\tfrac{1}{2}$ kg of flour. There is 1 kg 450 g in it. How much more flour will it hold?

11 From a piece of beef weighing 4·680 kg one piece weighing 2·755 kg and a smaller piece weighing 875 g were cut off. What was the weight of the piece that was left?

12 Two bottles of sauce together weigh 1·264 kg. Each bottle when empty weighs 365 g. How much sauce is there in each bottle?

13 Last week mother bought 750 g of mince. This week she bought a shoulder of mutton 2·3 times as heavy. What was the weight of the mutton?

14
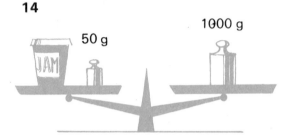
What is the weight of the pot of jam?

15 A tin of grapefruit juice weighs 542 g. How much juice is there in 6 tins?

16 A sack contained 50 kg of potatoes.
   (a) When 15·750 kg and 12·500 kg had been taken out, how many kg were left?
   (b) What is the value of the potatoes left at £0·36 a kg?

17 A bottle of sauce weighs 563 g. The bottle when empty weighs 308 g. How much sauce (in kg) is there in a dozen bottles?

18
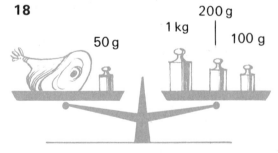
What is the weight of the leg of lamb?

19 A van carries 30 boxes of apples, each box weighing 50 kg. The van when empty weighs 2 460 kg.
   Find: (a) the weight of the apples
         (b) the weight of the loaded van
         (c) the value of the apples at 32p a kg.

20 A brick weighs 1½ kg. Find the weight of 250 bricks.

21 Tom is 5·445 kg heavier than Alan who weighs 36·380 kg. What is Tom's weight?

22 An empty box weighs 225 g. When 20 cakes of soap are put into it, it weighs 1·625 kg. What is the weight of a cake of soap?

23
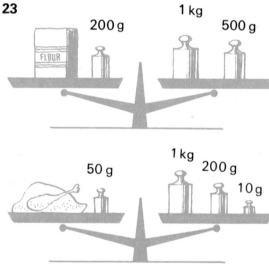
What is the weight of: (a) the flour? (b) the turkey?

24 Share 92·592 kg of sugar equally among 8 women. How much will each receive?

25 After selling 6·350 kg, 6·925 kg and 7·500 kg from a chest of tea a merchant had 8·500 kg left.
   (a) How much tea was in the chest at first?
   (b) What was the value of the tea left in the chest at 78p a kg?

Proportion

> Example:
> If 6 metres of cloth cost £3·90, how much will 8 metres cost?
> (a) First find the cost of 1 metre.
> (b) Now multiply by 8 to find the cost of 8 metres.
>
> 6 m cost £3·90
> so 1 m costs £3·90 ÷ 6 = £0·65
> 8 m cost £0·65 × 8 = £5·20

A
1. 12 Maths books cost £5·76. What will 7 books cost?
2. If 9 metres of material cost £6·48, how much will 11 metres cost?
3. If 4 litres of milk weigh 4·12 kg what is the weight of 9 litres?
4. My car can travel 192 km on 16 litres of petrol. How far on 20 litres?
5. If 5 metres of wire weigh 270 grammes, what would be the weight of 12 metres?
6. A man earns £27·30 for 6 days work. How much should he earn in 9 days?
7. In 5 hours a car covers a distance of 320 km. At this speed how far should it travel in 7 hours?
8. If 15 packets of grass seed weigh 750 grammes, what would 24 packets weigh?
9. When full 2 oil drums hold between them 560 litres of oil. How many litres would 9 drums of the same size hold?
10. An aeroplane travels 450 km in 30 minutes. At this speed how far would it travel in 45 minutes?

B
1. A length of 3·5 metres of curtain material costs £4·20. What would a length of 9 metres cost?
2. If 12 metres of lead pipe weigh 37·2 kg, what would 15 metres weigh?
3. In France I bought 9 metres of cotton for 30·60 francs. How much would I have paid for 20 metres?
4. I saw in a Paris shop a basket of apples weighing 12 kg. The price was 15·36 francs. How much would I have paid for a 10 kg basket?
5. A baker uses 100 kg of flour to make 125 kg of bread. How much bread should he get from 60 kg of flour?
6. A cyclist travels 750 metres in 3 minutes. At this speed how many kilometres would he travel in 20 minutes?
7. If 17 litres of alcohol weigh 13·6 kg, what is the weight of 30 litres?
8. A motor van does 100 km on 18 litres of petrol. How much petrol would it use to travel (a) 250 km, (b) 850 km?
9. If 3 girls earn between them £25·14 in a week, how much would 11 girls earn if they were paid at the same rate?

Buying and selling

A

1. A lady spent £6·75 on a dress and £4·95 on a hat. She had £2·48 left. How much had she at first?
2. I gave a £5 note to pay for 5 metres of cloth at 64p a metre and 4 metres of ribbon at 12½p a metre. How much change did I receive?
3. How much money does Dad need to buy 3 ties at 89p each and 2 shirts at £2·36 each?

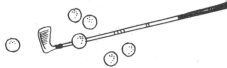

4. Dad bought a golf club and 6 balls for £7·10. The club cost £5·60. What was the price of a ball?
5. Mum bought a gigot of lamb weighing 1·500 kg. How much did she pay for it if the price was 68p a kg?
6. Mother went shopping. In her purse she had two £1 notes, a 50p coin, three 10p coins, four 5p coins and seven 2p coins. She spent £1·10 at the grocer's, 76p at the butcher's and 69p at the fruit shop. How much had she left?
7.

    From a roll of cloth measuring 30 metres a shop sold 1·5 m, 3·7 m, 4·3 m and 3·6 m.
    (a) How much cloth was sold?
    (b) What was the length of the piece that was left?

B

1. Find the total cost of:
   3 m of cloth at 48p a metre
   4 m of cotton at 65p a metre
   2 m of satin at 75p a metre
   5 m of tweed at £1·25 a metre.
2.

   (a) A man bought 2 kg of grapes and 3 kg of pears. What change did he receive from £1?
   (b) How much less than £2 would 3 kg of grapes and 5 kg of pears cost?
   (c) How much would I pay for 500 grammes of pears and 250 grammes of grapes?
3. The grocer's bill came to £1·15. The butcher's bill was 48p less. How much did the two bills amount to together?
4. A bus uses 34 litres of diesel oil every 100 kilometres. Every day it makes 2 return journeys between two towns 75 kilometres apart.
   (a) What distance does it travel every day?
   (b) How much diesel oil does it use each day?
   (c) What does the oil cost per day if the price is 7p a litre?

C 1

(a) How much would I pay for 9 bottles?
(b) If I bought 5 bottles how much change would I get from a £1 note?

2 What would 250 grammes cost at £1·28 a kilogramme?

3 An oil drum full of oil weighs 120 kg. The drum when empty weighs 12 kg.
(a) What is the weight of the oil?
(b) What is its value at £0·16 per kg?

4 A lady had a £5 note in her purse. She spent 24p less than £3. How much had she left?

5 I paid £1·04 for 10 kg of potatoes at $3\frac{1}{2}$p a kg and 3 kg of apples. What was the price of 1 kg of apples?

6 Jim spent £1·24. This was twice as much as Harry spent. How much did the two boys spend between them?

7

A fruiterer bought 6 cases of fruit, each weighing 9 kg, at 17p a kg. What was the total cost of the fruit?

8 A table and 4 chairs cost £41·50. The table cost £12·50. How much did each chair cost?

9 From a roll of material measuring 36·50 metres I cut 19·25 metres. What was the value of the piece that was left at £0·80 a metre?

D 1 A shopkeeper bought a dozen vases for £9·55. He sold them all at £1·15 each. How much profit did he make?

2 A merchant bought 40 metres of cloth for £23·50. He sold it all at £0·68 a metre. How much profit did he make?

3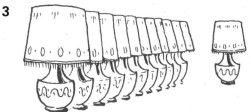

A dozen lampshades were bought at £1·10 each and sold at £1·65 each. How much profit was made?

4 A merchant bought 50 metres of material at £0·85 a metre. He sold it all at £1·30 a metre. How much profit did he make?

5 9 dozen tins of fruit were bought at 8p a tin and sold at $10\frac{1}{2}$p a tin. How much profit was made when all the tins were sold?

6 A shopkeeper bought $2\frac{1}{2}$ dozen pairs of gloves at £0·56 a pair. He sold them all at £0·80 a pair. How much profit did he make?

7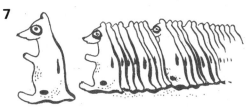

50 toys were bought at 45p each. All were sold at 60p each. How much profit was made?

8 3 dozen rugs were bought for £180. They were all sold at £5·85 each. How much profit was made?

9 During a sale a shop reduced its prices by 12p in the £. What was the sale price of a coat that was usually sold at £14·50?

Percentages

We often see notices like these. Let's find out what they mean.

Here are 100 marbles. 26 out of the 100 are coloured.

Another way of writing "26 out of 100" is $\frac{26}{100}$.

So we can say that $\frac{26}{100}$ of the marbles are coloured.

Another way of writing "26 out of 100", or $\frac{26}{100}$, is 26 **per cent**.
(**cent** is short for the Latin word **centum** which means one hundred.)

74 out of the 100 marbles are black

So 74 per cent are black.

Instead of the words **per cent** we often use the symbol %

So $\frac{26}{100}$ = 26 per cent = 26%; $\frac{74}{100}$ = 74 per cent = 74%.

Write in this way: $\frac{25}{100}$ $\frac{32}{100}$ $\frac{50}{100}$ $\frac{75}{100}$ $\frac{96}{100}$ $\frac{5}{100}$

Cut out three squares of graph paper, each divided into 100 small squares.

Each small square is $\frac{1}{100}$ or 1% of the large square

Colour them as shown and paste them into your exercise book.

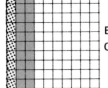
Black 10%
Colour 20%

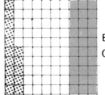
Black 15%
Colour 30%

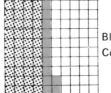
Black 40%
Colour 12%

It is easy to write percentages as fractions, since a percentage is just a fraction whose denominator is 100.

Example 1

$$10\% = \frac{\cancel{10}^1}{\cancel{100}_{10}} = \frac{1}{10}$$

Example 2

$$75\% = \frac{\cancel{75}^3}{\cancel{100}_4} = \frac{3}{4}$$

Note:

$$12\tfrac{1}{2}\% = \tfrac{1}{8} \qquad 33\tfrac{1}{3}\% = \tfrac{1}{3}$$

A Change these percentages to ordinary fractions in their simplest form:

	(a)	(b)	(c)	(d)	(e)	(f)
1	20%	60%	40%	25%	65%	$12\tfrac{1}{2}\%$
2	50%	80%	90%	75%	24%	$33\tfrac{1}{3}\%$
3	30%	70%	5%	35%	16%	$8\tfrac{1}{3}\%$

Example 4

What is 20% of £45?

$$20\% = \tfrac{20}{100} = \tfrac{1}{5}$$

$$\tfrac{1}{5} \text{ of } £45 = £9$$

Study Example 4 and then find:

4 (1) 50% of 30 eggs
   (2) 10% of 40 oranges
   (3) 30% of 50 pencils
   (4) 40% of 60 sweets
   (5) 25% of £24
   (6) 5% of £80
   (7) 15% of 40 boys
   (8) 75% of 20 girls
   (9) 60% of 35 men
  (10) 70% of 50 women

B Find:

	(a)	(b)	(c)	(d)
1	25% of £8	20% of 30 apples	35% of 40 pencils	15% of 100 boys
2	50% of £12	30% of 40 oranges	95% of 60 books	55% of 80 girls
3	75% of £16	60% of 60 pears	50% of 24 pens	32% of 50 men
4	50% of £10	80% of 50 plums	25% of 16 books	$12\tfrac{1}{2}\%$ of 40 women
5	100% of £3	45% of 80 bananas	75% of 20 pencils	$33\tfrac{1}{3}\%$ of 24 houses

C

1 John spends 75% of his money. What percentage has he left?

2 May had £6·40. She spent 30% of it. How much money had she left?

3 There are 36 pupils in my class. 25% are girls.
  (a) How many girls in my class?
  (b) How many boys?

4 There were 8 dozen eggs in a box. 50% were broken.
  (a) How many were broken?
  (b) What percentage was unbroken?

5 A man left £3 600. His son received 45% of it and his daughter received the remainder.
  (a) How much did the son get?
  (b) How much did the daughter get?
  (c) What percentage of the money did the daughter receive?

6 Dad's salary is £1 600 a year. He is to have a 5% increase. What will his salary be then?

> **Increase means "make bigger"**
> **Reduce means "make smaller"**

When the price of an article is increased, does the shopkeeper add something to the price or does he take something off?

What happens when the price is reduced?

Now do you see what the notices at the top of page 30 mean?

D Find the Sale Price of the following goods:

(1) **Chair**: 10% off
Usual Price   £10
Reduction   ?

Sale Price   ?

(2) **Table**: 5% off
Usual Price   £20
Reduction   ?

Sale Price   ?

(3) **Radiogram**: 20% off
Usual Price   £40
Reduction   ?

Sale Price   ?

(4) **Sideboard**: 25% off
Usual Price   £36
Reduction   ?

Sale Price   ?

(5) **Golf Clubs**: 10% off
Usual Price   £5
Reduction   ?

Sale Price   ?

(6) **Television**: 15% off
Usual Price   £60
Reduction   ?

Sale Price   ?

E

1. A man bought a car for £450. He sold it and made a profit of 10%. At what price did he sell the car?

2. Dad bought a car for £520. He sold it a year later and made a loss of 20%. At what price did he sell the car?

3. A dealer bought a bedroom suite for £160. At what price must he sell it to make a profit of 30%?

4. A merchant bought 3 articles at the following prices:
   (1) £2·40 (2) £5·16 (3) £7·36
   At what prices must he sell them to make a profit of 25% on each article?

5. Our house cost £3 500. We have now sold it at a profit of 45%. How much did we get for the house?

6. What is 75% of £9·36?

F A percentage is very like a decimal fraction.

> $0·75 = \frac{75}{100} = 75\%$

> $0·05 = \frac{5}{100} = 5\%$

Now write these decimal fractions in the same way as percentages:

0·65   0·25   0·95   0·04   0·01   0·45   0·5

Plans and scales

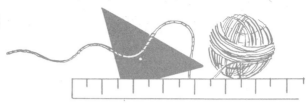

A 1 We have already learned that we can make a smaller length stand for a longer length.
In the 5 lines below 1 cm stands for 10 m (or 1 mm for 1 m)

We say that the scale is 1 cm to 10 m.

Measure the lines as carefully as you can in cm and mm and so find the longer lengths that the lines stand for.

(a) ——————————— (b) ——————————— (c) ———————————————

(d) ——————————————— (e) ———————————————

2 In the lines below the scale is 1 cm to 20 m.

First decide how many metres 1 mm stands for and then find the longer lengths represented by the lines.

(a) ——————————————— (b) ——————— (c) ———————————

(d) ——————————————————— (e) ———————————————

3 Here is the plan of a garden (that is, a "scale drawing" of a garden, looking at it from above—as if you were in an aeroplane, or on top of a high building.)

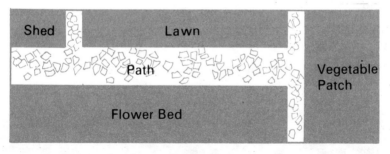

The scale of the plan is 1 cm to 4 m

Measure the following, and write them down in your exercise book.

The length and breadth of: (a) the garden (b) the shed (c) the lawn
(d) the flower bed (e) the vegetable patch.
(f) Measure the perimeter of the garden.

4 Use the scale to write down the real lengths of each of the above (in metres).

5 Draw this plan in your exercise book, and try to measure as carefully as you can with your ruler. Make sure your angles are right angles.

MATHS 4–C

6 If possible, perhaps you could use your rulers, metre strings and right angles, to chalk out in your playground the exact size of the garden, the shed, and so on.

7 Measure the length and breadth of your playground or school garden and draw a plan to scale.

8 If the scale of a plan is 1 cm to 10 cm, write down the scale length in centimetres of the following "real" measurements.
   (a) 50 cm   (b) 1 m   (c) 1·5 m   (d) 4 m   (e) 4·5 m

9 Here is the plan of a room. Draw it, but first measure the lengths and breadths, and also the width of the border.

   If the scale is 1 cm to 1 m, what are the real measurements of the room?

10 Measure the length and breadth of your classroom and draw a plan of it to the same scale.

B   In maps, lines of 1 cm can stand for 1 km, or 2 km or 6 km and so on.

1 Here are some road lengths on a map where the scale is 1 cm to 5 km. Measure each line, and then write down the distance it stands for:

   (a) ———————————————   (b) ———————————————
   (c) ——————————————————————   (d) ———————

2 If 1 cm stands for 10 km, draw lines to stand for:
   30 km   40 km   55 km   60 km   42 km.

3 If the scale is 1 cm to 20 m, draw lines to show:
   60 m   100 m   120 m   90 m   130 m.

4 Here is part of a map, showing some towns. It is drawn to the scale shown below the map.

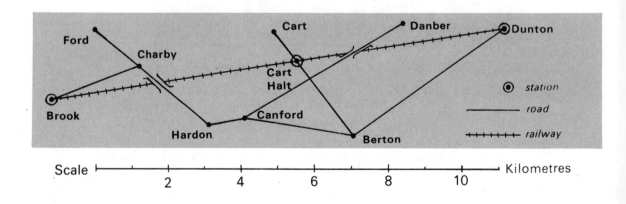

(a) How far is it from Brook to Charby?
(b) How far is it from Charby to Hardon?
(c) Is Hardon nearer to Charby than Canford?
(d) Which is farther from Berton and by how many km—Cart or Dunton?

5 Find out the distance from:
(a) Cart to Berton (b) Danber to Canford (c) Cart to Cart Halt
(d) Berton to Dunton (e) Ford to Canford (f) Brook to Dunton (by rail)

6 How far is Cart Halt from (a) Brook (b) Dunton?

7 Which is the shortest way from Brook to Dunton.
(a) by road (b) by train?

8 Find out the distance, if you were going by car, from Ford to Danber.

C   Toy dolls, cars, boats and trains are made much smaller in scale than real girls, or cars, or boats, or trains. In each of these we have a smaller scale.
Here are some pictures of toys, made smaller in size, with the scale shown below.
We say that these are **Scale Models** of the real things.

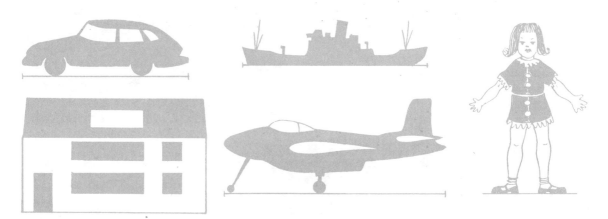

The scales are as follows:
Car: 1 cm to 1 m   **Ship:** 1 cm to 50 m   **Girl:** 1 cm to 30 cm
House: 1 cm to 4 m   **Aeroplane:** 1 cm to 2 m.

1 With your ruler measure the length of the car, the ship, the house and the aeroplane. Write down these lengths in a table like this:

	Model	Real	Model	Real	Model	Real	Model	Real
	Car	Car	Ship	Ship	Aeroplane	Aeroplane	House	House
Length								
Height								

Now use the scales shown below the drawings on the previous page to find the real lengths and enter them in your table.

2 If the doll is a model of a girl called Cindy, what is Cindy's real height in cm?

3 Measure the lengths, breadths or heights of your blackboard, the top of your desk or table, the classroom door, the classroom floor and so on.
Choose a good scale, and make scale drawings of them in your book.

D On maps, roads often wind in and out.
We can get a rough idea of distance between 2 places, A and B, by measuring with a ruler, if the road between them is fairly straight.
What, roughly, is the distance from A to B here?

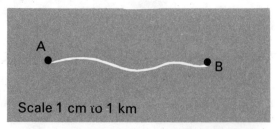
Scale 1 cm to 1 km

1 Sometimes we want to know distances more exactly.
We can use wire which is easily bent. Bend the wire round the curves from C to D. When you reach D, hold the wire there between finger and thumb, and straighten by pulling the end at C.
Now measure the length of the stretched out wire along a ruler, and show the length of the road in km.

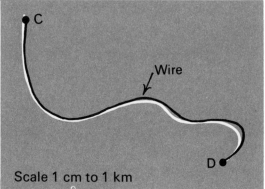
Scale 1 cm to 1 km

2 Note the scale and find the approximate distances from:

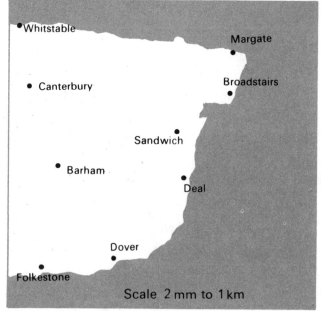

(a) Canterbury to Barham
(b) Folkestone to Canterbury
(c) Dover to Folkestone
(d) Whitstable to Deal
(e) Margate to Sandwich
(f) Broadstairs to Dover.

Scale 2 mm to 1 km

Making sure 1

A 1 By the main road the distance from Parkvale to Redstones is 8·5 km, but there is a short cut which saves 1·75 km. How far is it by the short cut?

2 When on holiday I bought a camera which cost £22·50. How much change did I get from three £10 notes?

3 What is the value of **n** in each of the following equations?
   (a) n + 3 = 12   (b) n − 5 = 9   (c) $\frac{1}{5}$ of n = 8   (d) $\frac{1}{10}$ of n = 18

4 Write the following percentages as fractions:
   40%   70%   75%   25%   5%   50%

5 Change the following decimal fractions to percentages:
   0·5   0·6   0·25   0·75   0·46   0·85

6 Find: (a) 50% of £16   (b) 25% of £36   (c) 75% of 24   (d) 80% of 40.

B 1 A car uses 9 litres of petrol on a journey of 108 km. How far does it travel on (a) 1 litre   (b) 8 litres   (c) 12 litres   (d) 19 litres?

2 If a road map is drawn to the scale of 1 cm to 10 km, what would be the real distances of these lines on the map:
   (a) 4·5 cm   (b) 5·2 cm   (c) 6·3 cm   (d) 3·8 cm   (e) 2·4 cm?

3 If 1 cm stands for 8 km, draw lines to stand for:
   (a) 16 km   (b) 24 km   (c) 28 km   (d) 44 km   (e) 36 km.

4 A carpet is 4 metres long and $3\frac{1}{2}$ metres wide. How many square metres of floor will it cover?

5 These lines stand for the lengths of 5 of the world's great rivers. The scale used is 1 cm to 1 000 km. Find the approximate length of each river.

   Danube   ————————
   Congo   ——————————————
   Nile   ——————————————————
   Amazon   ———————————————————
   Mississippi ————————————————————

6 A plane left Gatport at 17.15 and landed at Lemton at 18.35, flying at an average speed of 570 km an hour.
   (a) How long did the flight take?   (b) How far is it from Gatport to Lemton?

Angles and tiles

A   In 1 hour, the minute hand of a clock makes 1 complete turn (or revolution).

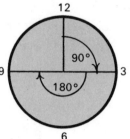

For easy measuring, we divide this complete turn
into 360 equal parts.
Each part is called a **DEGREE**.
A Right Angle ($\frac{1}{4}$ turn) measures 90 degrees (90°).
A Straight Angle ($\frac{1}{2}$ turn) measures 180 degrees (180°).

1 If we start at 12 o'clock, through how many degrees does the hour hand turn to reach (o'clock):
  (a) 1   (b) 3   (c) 4   (d) 6   (e) 8   (f) 9   (g) 10   (h) 11?

2 What number will the minute hand point to, if it turns through:
  (a) 30°   (b) 60°   (c) 90°   (d) 150°   (e) 180°   (f) 210°   (g) 270°   (h) 300°?

B   The diagrams below show wheels with their spokes spaced evenly.

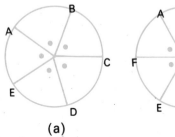
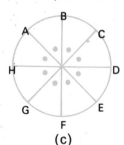

  (a)   (b)   (c)   (d)

1 What fraction of 1 complete turn is the angle between each spoke?
2 What is the angle between the spokes in each diagram?
3 Trace each figure, and join A to B, B to C, C to D and so on.
   Each shape is called a Regular Polygon. (Polygon means "many-sided", and regular means all the sides are equal.)

> Shape (a) has 5 equal sides, and is called a regular **PENTAGON**
> Shape (b) has 6 equal sides, and is called a regular **HEXAGON**
> Shape (c) has 8 equal sides, and is called a regular **OCTAGON**
> Shape (d) has 12 equal sides, and is called a regular **DODECAGON**

4 Draw up a table, like this, for the polygons (a), (b), (c) and (d) on the previous page.

Name	Sides	Angle at Centre	Fraction of 360°
Pentagon	5	72°	$\frac{1}{5}$
	6		
	8		
	12		

C **1** Use tracing paper to trace these diagrams.

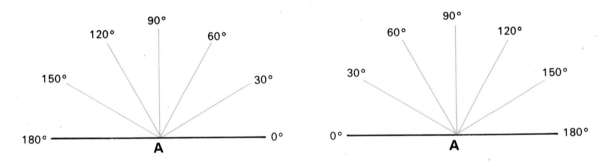

We call each tracing a protractor. A protractor is used for measuring the size of angles.

2 Use the first tracing to measure the first 4 angles, and the second tracing to measure the other 3 angles. Choose the proper tracing to measure other angles. When measuring, place the base line along an arm of the angle and put the point A on the vertex.

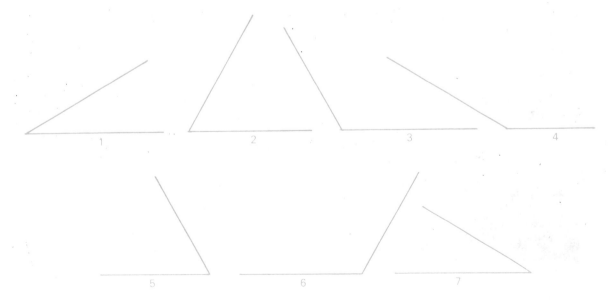

3 Now draw some angles in your exercise book, and use your protractor to estimate their size. Ask your neighbour to check your estimate and you check his.

We can use our tracings, or protractors, to draw angles of various sizes.

Draw a line XY 5 cm long.
Place the base line of your protractor along XY, so that A is on top of X.
Now stick a pin, or other sharp point, through the end of the line marked 60°.
Join your point to X.
You have now drawn an angle of 60°.

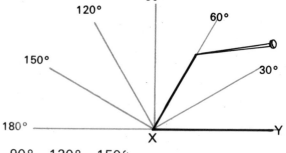

D 1 Draw in this way angles of: 30° 60° 90° 120° 150°

2 Draw by estimating as close as you can angles of: 15° 45° 75° 105° 135° 165°

E Tiles

1 Here are 4 square tiles fitted together. What is the sum of the angles marked?

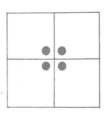

2 Fit some cardboard rectangles of the same size in various ways to cover part of your desk without leaving gaps. Here is one way.

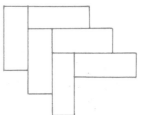

3 If you have some cardboard, or plastic, shapes—triangles, parallelograms, kites, hexagons, and so on, try to cover space with them in the same way without leaving any gaps. If you can do this, what do you notice about the sum of the angles round each point where the tiles meet?

4 If you have no shapes you can make some in this way. Trace the shapes below, place on cardboard and stick a pin through each vertex of the tracing. Join up the pin holes. Cut out each shape, draw round it on cardboard, and so make several congruent shapes. Try to cover space with them without leaving gaps.

5 Here is part of a tiling with regular hexagons. The angles at a meeting point are marked with a dot.
What is the sum of the angles at a meeting point?
What is the size of each angle?
What then is the size of each angle of a regular hexagon?
Where is this tiling often seen?

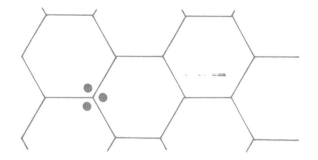

6 Draw some congruent shapes on squared paper to make a tiling, and find the size of each angle round a meeting point.

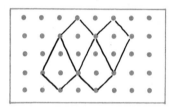

7 If you have a nail-board, use elastic bands to make a tiling with various congruent shapes. Here is one.

8 Place some round counters (or coins) on your desk to touch each other. Are there any gaps between them? Could you then make a tiling with circles?

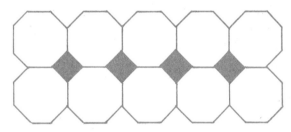

9 Find out some shapes which will leave gaps when fitted together.
Here is one.
Do you know its name?

10 On squared paper, try to draw some diagrams, like these below.

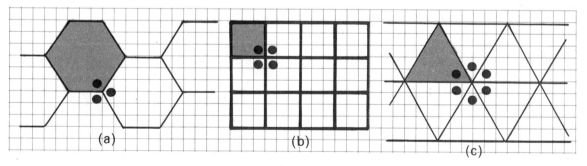

(a) 3 tiles meet at each point—Size of each angle = 120°
(b) 4 tiles meet at each point—Size of each angle = 90°
(c) 6 tiles meet at each point—Size of each angle = ?°
(d) Now make other tiles meet at a point, and find the size of each angle at the meeting point.

41

Circle patterns

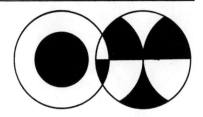

A

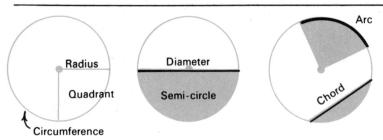

Here are some parts of a circle.
The dot is the centre.

1   Now copy and complete these sentences.

   (a) The circumference is the PE......R of a circle.
   (b) A radius is the distance from the centre to the CIR.........E.
   (c) The diameter is twice the R....S.
   (d) The diameter divides a circle into 2 equal S........ES.
   (e) A quarter of a circle is called a Q......T.
   (f) An arc is part of the CIR.........E.
   (g) A C...D joins any 2 points on the circumference.

2 (a) Use your compasses to draw 3 circles with the same centre and with radii: 3 cm, 3·5 cm, 4 cm. Colour in between circles to make a target.
   (b) Draw 3 semicircles, and 3 quadrants with these radii.
   (c) What is the diameter of each circle?

B   Use compasses to draw these designs. Look carefully to find the centres.

(a) 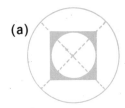 (b) 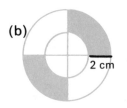 (c)  (d)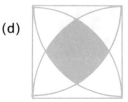

Start with a 3 cm square | Outer circle has a radius 4 cm | 4 cm square | 4 cm square The centre of each quadrant is a corner of the square.

1 In designs (a) and (b), what is the radius of the inner circle? What is the diameter?

2 What is the radius of the quadrant in (d)?

3 Now draw some designs of your own, using circles, semi-circles and quadrants.

C In Book 3 we drew hexagons, using the radius of a circle to mark off the circumference of the circle into 6 equal arcs.
Draw a circle of radius 3 cm, and mark off a hexagon.
What size is each side?
Now copy the designs below, using a **radius of 3 cm for each outer circle**.
In each case mark off the circle as above.

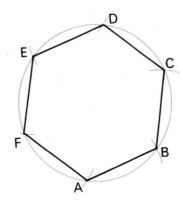

 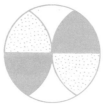  

D **1** Here are some traffic signs. Draw them (rather bigger). Find out the actual colours and colour your signs in the same way.

  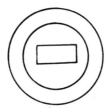  

Speed Limit • End of speed limit • No entry • No stopping • Stop and give way

2 Find out what the following signs look like, draw them and put in the right colours.

(a) No waiting
(b) Stop—children crossing
(c) Play Street
(d) Meter zone
(e) Roundabout
(f) Traffic lights

Drawing curves

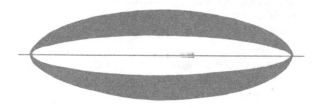

A The Ellipse

If you tilt a tumbler half full of water at an angle and look down, the surface of the water forms a curved shape called an **ellipse**.
The top part of a tea cup or a bucket looks like an ellipse when viewed from an angle, though it is a circle if looked at from above. This helps us to draw such objects.
Try to draw these.

A good ellipse is difficult to draw, but you could draw one in this way:

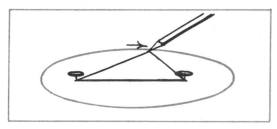

Stick two drawing pins in a piece of paper on a wooden surface. Place them 5 or 6 cm apart. Put a loop of string over the pins. Put the point of a pencil in the loop, keep it tight, and as you move the pencil you will draw an ellipse.

Draw some more ellipses by changing the distance between the pins.
Draw in any lines of symmetry you can find.
You could use this idea with 3 pieces of wood to mark out an oval flower bed in a garden, using one piece like the pencil.
You could use 2 sticks and a piece of chalk to draw a large ellipse in your playground.

The earth goes round the sun in a path which is an ellipse.

B The Parabola

Here are 2 U-shaped curves, each with a line of symmetry down the middle.

Each curve is called a **parabola**.
Try to draw one or two such curves freehand to make these sketches.

Here are some examples of parabolas, or paths which follow a parabola.

The main cable of a suspension bridge

Water from a hose

Boy throwing a ball

Can you think of any more "parabolic" paths, or shapes showing a parabola?

C Curves from straight lines

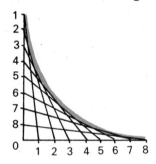

1 Draw 2 straight lines, each 8 cm long at right angles to each other and mark off both lines in cm. Number the points as shown.
Draw a straight line from point 1 on the vertical line to point 1 on the horizontal line.
Now join the 2 points numbered 2 and so on.
When you have finished you will see a curve, though you have only drawn straight lines.

2 Do the same again but this time draw the vertical line at the other end of the horizontal line.

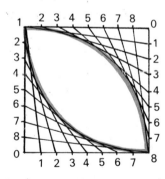

3 We can put these shapes together, like this, by using an 8 cm square marked in cm as shown.

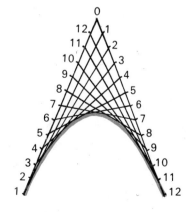

4 Draw two straight lines, each 12 cm long, to make an angle.
Mark off in cm and number as shown.
Now join up the points as before.
What shape is the curve?

45

Mathematical sentences or equations

A Look at these equations.

$n + 3 = 7$
We can find the value of n which will make the equation a true statement in this way:

Subtract 3 from each side.
$n + 3 - 3 = 7 - 3$
So $n = 4$

$n - 2 = 6$
Add 2 to each side.
$n - 2 + 2 = 6 + 2$
So $n = 8$

Now, using the ways shown above, find the value of **n** in the following equations:

1 $n + 2 = 5$
2 $n + 6 = 20$
3 $n + 4 = 10$
4 $n - 4 = 6$
5 $n - 7 = 18$
6 $n - 3 = 5$
7 $n + 5 = 15$
8 $n - 8 = 22$
9 $n + 6 = 8$
10 $n - 2 = 13$
11 $n + 9 = 20$
12 $n - 5 = 30$

B We can compare equations with a pair of scales.

1 kg 1 000 g

It is clear that if we want the scales to balance we must have the same weight on each side.

(a) If we doubled the weight on one side, what would we have to do to the other side to keep the balance?
(b) If we halved the weight on one side, what would we do to keep the balance?
(c) If we multiplied the weight on one side by 3, what would we have to multiply the other side by to keep the balance?
(d) If we divided the weight on each side by 4, would the scales balance?

Here is another equation:
$2a = 8$
Divide each side by 2.
So $a = 4$

We know that $2a$ means "**two times a**" or "**a multiplied by 2**".
Let's check our answer.
$2a = 2 \times 4 = 8$

46

1 Now find the value of **a** in these equations. Always check your answer.

(1) $2a = 6$ (3) $4a = 12$ (5) $6a = 18$ (7) $9a = 36$
(2) $3a = 12$ (4) $5a = 20$ (6) $8a = 40$ (8) $10a = 90$

We know that $\frac{n}{2}$ means $\frac{1}{2}$ of **n**, and $\frac{2}{3}$ of **y** = $\frac{2y}{3}$

Look at these equations:

$\frac{1}{2}$ of n = 6
Multiply both sides by 2
So n = 12

$\frac{1}{3}$ of y = 6
Multiply both sides by 3
So y = 18

2 Find the value of **n** in these equations and check your answer.

(1) $\frac{1}{2}$ of n = 10   (3) $\frac{1}{3}$ of n = 3   (5) $\frac{1}{6}$ of n = 3   (7) $\frac{1}{7}$ of n = 4
(2) $\frac{1}{4}$ of n = 2   (4) $\frac{1}{5}$ of n = 4   (6) $\frac{1}{8}$ of n = 4   (8) $\frac{1}{10}$ of n = 5

3 Work these out:

(a) $\frac{1}{2} \times \frac{2}{1}$ (b) $\frac{1}{4} \times \frac{4}{1}$ (c) $\frac{1}{3} \times \frac{3}{1}$ (d) $\frac{5}{1} \times \frac{1}{5}$ (e) $\frac{2}{3} \times \frac{3}{2}$
(f) $\frac{4}{5} \times \frac{5}{4}$ (g) $\frac{5}{6} \times \frac{6}{5}$ (h) $\frac{4}{3} \times \frac{3}{4}$ (i) $\frac{3}{8} \times \frac{8}{3}$ (j) $\frac{2}{5} \times \frac{5}{2}$

Notice that 2, 3, 4 can be written $\frac{2}{1}$ $\frac{3}{1}$ $\frac{4}{1}$

What did you notice about all your answers?

What did you notice about each pair of fractions?

When we get two fractions like these, where one of them is "the other way round" (or "upside down"), we say that one fraction is the **INVERSE** of the other.

When we multiply a fraction by its inverse, the result is 1.

For example, $\frac{2}{3} \times \frac{3}{2} = 1$.

Notice that, to find the inverse of a whole number, we divide 1 by the whole number.

Thus the inverse of 2 is $\frac{1}{2}$, the inverse of 4 = $\frac{1}{4}$, and so on.

4 Write down the inverse of the following:

$\frac{1}{3}$ $\frac{1}{6}$ $\frac{1}{5}$ $\frac{2}{3}$ $\frac{3}{4}$ $\frac{7}{8}$ $\frac{5}{4}$ $\frac{11}{4}$

4   5   8 $\frac{8}{3}$ $\frac{9}{7}$ $\frac{3}{11}$ $\frac{6}{5}$ $\frac{7}{12}$

Remember when you multiply a fraction by its inverse, the answer is always 1.

This leads us to an easy way of finding the value of **n** in equations such as $\frac{2}{3}$ of n = 6, and $\frac{4}{5}$ of n = 8.

$\frac{2}{3}$ of n = 6

The inverse of $\frac{2}{3}$ is $\frac{3}{2}$

Multiply each side by $\frac{3}{2}$

So n = 6 × $\frac{3}{2}$ = 9

$\frac{4}{5}$ of n = 8

The inverse of $\frac{4}{5}$ is $\frac{5}{4}$

Multiply each side by $\frac{5}{4}$

So n = 8 × $\frac{5}{4}$ = 10

5 Now find the value of **n** in these equations:

(1) $\frac{3}{4}$ of n = 9   (5) $\frac{3}{4}$ of n = 12   (9) $\frac{3}{10}$ of n = 9   (13) $\frac{4}{7}$ of n = 20

(2) $\frac{2}{3}$ of n = 10   (6) $\frac{3}{8}$ of n = 6   (10) $\frac{5}{9}$ of n = 15   (14) $\frac{5}{9}$ of n = 25

(3) $\frac{1}{2}$ of n = 9   (7) $\frac{7}{8}$ of n = 14   (11) $\frac{3}{7}$ of n = 12   (15) $\frac{3}{10}$ of n = 27

(4) $\frac{2}{5}$ of n = 8   (8) $\frac{5}{6}$ of n = 10   (12) $\frac{9}{10}$ of n = 18   (16) $\frac{7}{12}$ of n = 21

When we find the value of **n**, or **a**, or **b**, or **y** in an equation, we say that we solve the equation, just as a crime is solved when the unknown criminal is found.

6 Now solve these equations:

(1) 5a = 15   (4) y − 6 = 4   (7) 3b = 12   (10) a + 10 = 19

(2) $\frac{1}{2}$ of a = 6   (5) n + 5 = 12   (8) $\frac{1}{4}$ of b = 4   (11) $\frac{5}{6}$ of n = 15

(3) $\frac{1}{3}$ of y = 5   (6) b − 8 = 4   (9) $\frac{3}{4}$ of n = 9   (12) $\frac{3}{8}$ of y = 9

C Make up an equation about each of the following and solve it. The first one is done for you.

1 I think of a number. When I double it, the answer is 10.
What number did I think of?
Call the unknown number **n** (You can, of course, call it **a, b, c** or any other letter).

2n = 10   So n = 10 × $\frac{1}{2}$ = 5

2 $\frac{1}{6}$ of the boys in a class were red-haired. If there were 5 boys with red hair, what was the number of boys in the class?

3 Mary spent $\frac{2}{3}$ of her pocket money. If she spent 30p, how much pocket money had she?

4 $\frac{3}{4}$ of the number of passengers in a bus were downstairs. If 15 passengers were downstairs, how many were on the bus?

5 $\frac{4}{5}$ of a sum of money amounts to £8. What is the sum of money?

6 2 out of every 7 of the boys in a class have bikes. If 6 boys have bikes, how many are in the class?

7 John cycles $\frac{5}{8}$ of the way between Carrick and Oakvale. If he cycles 15 km, how far is it from Carrick to Oakvale?

8 A bath is $\frac{7}{10}$ full and contains 140 litres of water.
How many litres of water can the bath hold when full?

Indices

A You remember we found a short way of writing square numbers.

$4 = 2 \times 2 = 2^2$ $9 = 3 \times 3 = 3^2$ $16 = 4 \times 4 = 4^2$

2^2 is read as "two squared"

1 Write down the next four square numbers in this way.
2 What is 9^2 10^2 12^2 20^2 50^2 ?

B Some numbers are Cubic numbers.
Here is a cube formed from 8 blobs of plasticene joined by sticks.

$8 = 2 \times 2 \times 2$, and we can write this in "shorthand" as 2^3 (two cubed).

How many twos are multiplied together?

$$27 = 3 \times 3 \times 3 = 3^3$$

1 What are the next two cubic numbers?
2 What is   (a) 6^3 (b) 10^3?
3 What do you think 3^4 and 2^5 would mean?
4 Work out the numbers 3^5 and 2^4.

C Note that $16 = 4 \times 4 = 4^2$

Here we are counting in fours and we say that the Base is 4. The small 2 above is called the **INDEX** or **POWER**. (16 is "4 to the power of 2").

Also $16 = 2 \times 2 \times 2 \times 2 = 2^4$

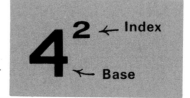

Here we are counting in twos. The Base is 2, and the Index is 4.
Write in this way:

(a) 8 with base 2
(b) 27 with base 3
(c) 32 with base 2
(d) 64 with base 4
(e) 64 with base 2
(f) 25 with base 5
(g) 49 with base 7
(h) 81 with base 3
(i) 64 with base 8

MATHS 4–D

D Here we see 32 expressed in two ways:

Note $32 = 2 \times 16 = 2 \times 2 \times 8 = 2 \times 2 \times 2 \times 4 = 2 \times 2 \times 2 \times 2 \times 2 = 2^5$
$32 = 4 \times 8 = 2^2 \times 2^3$

Can you see a way of finding 4×8 expressed in this way with the base 2 and only one index?
If two numbers are multiplied together and are given in the same base, we can **add** the indices.

Thus $4 \times 8 = 2^2 \times 2^3 = 2^5$ (2 to the fifth power).

1 Now show these with a single base:

(a) $2^3 \times 2^4$
(b) $3^2 \times 3^3$
(c) $5^2 \times 5^2$
(d) $7^2 \times 7^4$
(e) $10^3 \times 10^2$
(f) $2^2 \times 2^2 \times 2^2$
(g) $2^2 \times 2^3 \times 2^4$
(h) $3^3 \times 3^4 \times 3^5$
(i) $5 \times 5^2 \times 5^3$

2 Now show these with a single base:

(a) 4×4 (base 2)
(b) 9×27 (base 3)
(c) 25×25 (base 5)
(d) 8×8 (base 2)
(e) 27×27 (base 3)
(f) 16×16 (base 4)

Notice: $32 = 2^5$
$16 = 2^4$
$8 = 2^3$
$4 = 2^2$
$2 = 2^1$

E We can use indices for division. Look at these 3 ways of dividing 32 by 4

$$\frac{\cancel{32}^8}{\cancel{4}^1} = 8 \qquad \frac{32}{4} = \frac{2^1 \times 2^1 \times 2 \times 2 \times 2}{2^1 \times 2^1} = 2^3 = 8 \qquad \frac{32}{4} = \frac{2^5}{2^2} = 2^3$$

Can you see the quick way of finding the answer (using base 2)?

$\frac{2^5}{2^2} = 2^3$ 	We take the index of the denominator from the index of the numerator $(5 - 2)$ **as long as the number base is the same.**

In the same way $\frac{27}{9} = \frac{3^3}{3^2} = 3^1 = 3$

1 Now try these in this way:

(a) $2^6 \div 2^4$
(b) $3^7 \div 3^4$
(c) $4^5 \div 4^3$
(d) $5^4 \div 5^2$
(e) $7^4 \div 7^3$
(f) $10^5 \div 10^2$
(g) $(2^4 \times 2^3) \div 2^5$
(h) $(3^3 \times 3^5) \div 3^4$

2 Try these (but first show each number in the base given).

(a) 16 ÷ 4 (base 2) (c) 64 ÷ 8 (base 2)   (e) (81 × 9) ÷ 27 (base 3)
(b) 81 ÷ 9 (base 3) (d) 128 ÷ 16 (base 2) (f) (27 × 27) ÷ 81 (base 3)

F Here is a table of numbers showing the index of each to the base 2.

Number	2	4	8	16	32	64	128	256
Index	1	2	3	4	5	6	7	8

Thus $128 = 2^7$ (2 to the seventh power).

$16 \times 4 = 64$

Under 16 and 4, we find the indices 4 and 2.

If we add these (4 + 2) we get 6, and above the index 6 we find 64.

What we are really doing is this:

$16 \times 4 = 2^4 \times 2^2 = 2^6 = 64$

1 Multiply in this way:

(a) 8 × 16   (b) 4 × 64   (c) 32 × 4   (d) 128 × 2

2 Now copy the table into your exercise book and by doubling the top line each time, make it up to 2^{10} and find:

(a) 64 × 8   (b) 64 × 16   (c) 32 × 64   (d) 256 × 16

G We can use this idea to make a slide rule to multiply and divide numbers which can be written as "powers of 2".

Copy a table like this neatly into your exercise book:

A	1	2	4	8	16	32	64	128	256

Now copy the same table on to a strip of paper or cardboard whose top edge is a straight line. But call this second table strip B.

1 Now slide strip B along under table A so that the 8 on strip B is under the 1 on table A, like this:

A	1	2	4	8	16	32	64	128	256			
			B	1	2	4	8	16	32	64	128	256

We can now multiply each number in table A by 8.
The answer is shown on strip B under each number multiplied by 8.

Thus $2 \times 8 = 16$ $4 \times 8 = 32$ $32 \times 8 = 256$

In the same way we can put the 2 on strip B under the 1 on table A and you can now multiply by 2,

Now try these on your "slide rule".

(a) 4×16 (b) 4×64 (c) 16×8 (d) 16×16 (e) 32×8

You can keep on doubling the numbers on each scale and make up a longer "slide-rule".
You could also use two rectangular strips of cardboard.

2 You can use your "slide-rule" for division.

In the slide-rule drawing shown,

$128 \div 16 = 8$ (shown below the 1 of table A) and $32 \div 4 = 8$.

Use your slide-rule in this way to find:

(a) $64 \div 16$ (b) $128 \div 4$ (c) $256 \div 8$ (d) $256 \div 64$ (e) $128 \div 32$

3 In the same way you can make two more strips to multiply and divide numbers which can be written as "powers of 3".

Your strips will show the numbers 1, 3, 9, 27, 81, 243 . . . and so on.

Make up questions for your neighbour to answer on his "power of 3" slide-rule.

A	1	3	9	27	81	243			
B	1	3	9						

Counting bases

A We usually count in tens, since we have 10 fingers. Long ago some people used to count in twos (2 hands), and in fours (2 hands and 2 feet). When we count in tens, we use the numbers (or digits)—0, 1, 2, 3, 4, 5, 6, 7, 8, 9. "Digit" means "finger". When we come to ten, we use the digits 0 and 1. When counting in 10's we say our **counting base** is 10, or that we are using a "tens" count.

TEN = 10

Here are some ways of showing a **BASE 10** number, say 243.

(a)
ABACUS		
100	10	1
2	4	3

(b)
MONEY		
100p	10p	1p
2	4	3

1 In drawing (a): 243 = (100 × 2) + (10 × 4) + (1 × 3) = 200 + 40 + 3.
  In drawing (b): 243 means (100p × 2) + (10p × 4) + (1p × 3) = 200p + 40p + 3p
    = 243 p = £2·43.

2 (a) Draw some pictures to show what these numbers mean: 123, 234, 324, 235, £1·25, 123p.
  (b) Write down what each means in the way shown in 1 above.

3 Write down some numbers of your own, say what each means, and draw a picture of each.

B There is another way of showing the "base 10" count.

$100 = 10 \times 10 = 10^2$	10	1

or

100 cigarettes	10 cigs	1
100	10	

Suppose we have 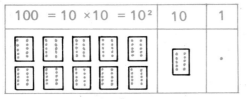 standing for 100, ▨ standing for 10 and ● standing for 1.

1 What numbers would these pictures show?

2 Draw pictures, like these, on squared paper, to show these "base 10" numbers:

    112   134   243   322   342   553.

(In each case, write down what each digit means.)

C When mother buys eggs, she usually buys them in sixes or twelves (dozens). We say the eggs are counted in **counting base** 6, or **counting base** 12

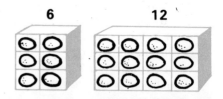

1 Write down the names of some things which are usually sold, (or counted), in these units:

    1's   2's   5's   6's   10's   12's

The pictures at the top of this chapter should help you

2 Let us try to count in SIXES, instead of TENS as if we had 3 fingers on each hand, instead of 5 fingers on each hand. Our counting base is now 6. We are counting in sixes.

We can show the base we are counting in like this

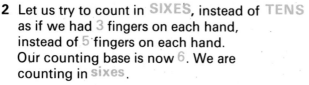

[We read 243_6 as "two, four three".]
When we count in 6's we use the digits—
0, 1, 2, 3, 4, 5.

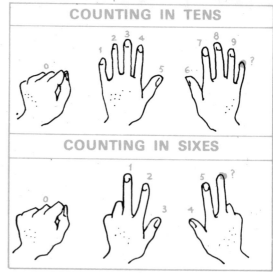

3 Here is a drawing showing the start of a base 6 count.

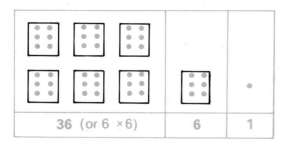

Note that 36 = 6 lots of 6, in the same way as 100 = 10 lots of 10.

54

If we let ◯ stand for 36 eggs, ◯ for 6 eggs, and ○ for 1 egg, what numbers do these pictures show:

(a) (b) (c)

(d)

4 Here is the start of the "tens count", and of the "sixes count".

(a)
	←	× 10	
10²	10	1	
100	10	1	BASE 10
2	4	3	

(b)
	←	× 6	
6²	6	1	
36	6	1	BASE 6
2	4	3	

From right to left, we multiply by 10 each time.

From right to left, we multiply by 6 each time.

(1) What will be the next number in the place to the left in base 10 ?

| ? | 100 | 10 | 1 |

(2) What will be the number in the next place to the left in base 6 ?

| ? | 36 | 6 | 1 |

(3) 243_{10} from table (a) means: $(100 \times 2) + (10 \times 4) + (1 \times 3) = 200 + 40 + 3$
   243_6 from table (b) means: $(36 \times 2) + (6 \times 4) + (1 \times 3) = 72 + 24 + 3 = 99_{10}$

So $243_6 = 99_{10}$

We read 243 as "two, four, three"— and **not "two hundred and forty three"**.

(4) Now show what each of these base 6 numbers are in base 10:
   123_6 221_6 145_6 301_6 230_6 500_6

5 Here is a way of changing numbers from base 10 count to base 6 count. Divide each time by 6 (the base), until your final answer is less than 6. The base 6 answer is read from * to the top right as shown

Change 243_{10} to base 6

```
6 | 243
6 |  40  R  3  ↑
6 |   6  R  4  |
  |   1  R  0  |
      *
```

$243_{10} = 1043_6$

```
10 | 243
10 |  24  R  3  ↑
   |   2  R  4  |
       *
```

We get 243 again!

Perhaps you can see how this works by looking at 243 (base 10). What happens when we keep dividing by 10, until our final answer is less than 10?

55

We can check our answer, like this:

[Note that 216 = 36 × 6.]

216	36	6	1
1	0	4	3

1043_6 = (216 × 1) + (36 × 0) + (6 × 4) + (1 × 3)
     = 216 + 0 + 24 + 3 = 243_{10}

Change, in this way, these base 10 numbers to base 6 numbers, and check your answers:
   247   328   412   502   360   720.

D We can count in threes. It is easy to pick up 3 eggs in one hand, so at one time people spoke about "2 handfuls of eggs" (6 eggs). Warehouses often sell things in "quarter-dozens" (threes):
   When we count in threes (**or base** 3), we use only the digits—0, 1, 2.
   221_3 is read as "**two, two, one** base three".

1 Here are pictures showing numbers counted in threes.

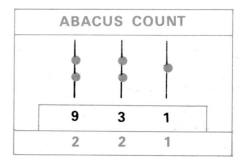

121_3 = (9 × 1) + (3 × 2) + (1 × 1) = 9 + 6 + 1 = 16_{10}
221_3 = (9 × 2) + (3 × 2) + (1 × 1) = 18 + 6 + 1 = 25_{10}

What will these base 3 numbers be in base 10:
   112   212   102   201   111   220   210?

2 (a)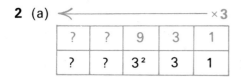

?	?	9	3	1
?	?	3^2	3	1

We multiply by 3 each time from right to left. What will be the next two numbers to the left?

(b) Copy the drawing, fill in the numbers and find out what 1221_3 and 11221_3 are in base 10. Write down some four or five digit numbers of your own in base 3 (remember to use only the digits 0, 1, 2) and change them to base 10 numbers.

(c) Here is a dot pattern for base 3.

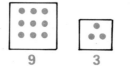

If ■ stands for 9, ■ stands for 3, and ● stands for 1
what would these pictures stand for:

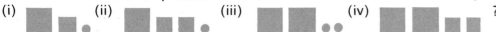

(i) ■ ■ ●   (ii) ■ ■ ■ ●   (iii) ■ ■ ● ●   (iv) ■ ■ ■ ■ ?

3 Changing base 10 numbers to base 3 numbers.
Let us change 25_{10} and 27_{10} to base 3 numbers.
In each case we divide by 3 as often as we can until the final answer is less than 3. (Remainders must be shown each time) We read from the final answer (∗) as shown and then upwards.

(a)
```
3 | 25
3 |  8 R 1 ↑
  |  2 R 2
  ∗
```
$25_{10} = 221_3$

(b)
```
3 | 27
3 |  9 R 0 ↑
3 |  3 R 0
  |  1 R 0
  ∗
```
$27_{10} = 1000_3$

We can check our answers in this way:

BASE 3	81	27	9	3	1		BASE 10
221			2	2	1	=	(9 × 2) + (3 × 2) + (1 × 1) = 18 + 6 + 1 = 25
1000		1	0	0	0		(27 × 1) + (9 × 0) + (3 × 0) + (1 × 0) = 27

∗Remember we say "two, two, one" and "one, zero, zero, zero".

(a) Write these base 10 numbers in base 3:
    17   20   26   48   62   90   100.

(b) Write these base 3 numbers in base 10:
    121   222   100   1 021   1 212   1 111

4 (a) What is the difference between $1\,022_3$ and 36_{10}?
   (b) What would be the base 10 answer to $[212_3 + 27_{10}]$?
   (c) What would be the base 3 answer to $[202_3 \div 10_{10}]$?

Try to make an abacus for base 3, and another for base 6.

Making sure 2

A 1 Which is the bigger: (a) 75% or 7·5 (b) 52% or 0·5 (c) £1·05 or 98p (d) 106p or £1·10?
   2 Which is the bigger and by how much: 0·5 of £100 or 0·75 of £60?
   3 What is the average height of 3 boys who are 1·64 m, 1·28 m and 1·43 m tall?
   4 A man earns £20 for 8 days' work. How much should he earn in
       (a) 1 day   (b) 3 days   (c) 10 days?
   5 Here is the timetable for 2 trains travelling from Aberdeen to London.

Aberdeen	Perth	Edinburgh	London
10.20	12.15	14.00	20.30
12.05	14.16	16.00	21.58

   (1) How long does each train take to go from Aberdeen to:
       (a) Perth   (b) Edinburgh   (c) London?
   (2) How long does each train take to go from Perth to:
       (a) Edinburgh   (b) London?
   (3) Copy this timetable but instead of 24-hour clock times, put in ordinary
       clock times (am and pm).

B 1 A square has sides of 6·5 cm. Find (a) the perimeter of the square
      (b) the area of the square.
   2 A road is divided into 4 traffic lanes, each 4·65 m wide. How wide is the road?
   3 Another road, 14·40 m wide, is divided into 3 traffic lanes of equal width.
      How wide is each lane?
   4 My watch loses 4 minutes every 6 hours. If I set it right at 9 am on Monday,
      what time will it show at 9 pm on Tuesday?
   5 A motorist travels 285 km in 5 hours. Travelling at this speed, how far should he
      travel in   (a) 1 hour   (b) 3 hours   (c) 7 hours?
   6 Find: (a) $6^2 + 4^2 - 2^2$ (b) $5^2 + 3^2 - 4^2$ (c) $8^2 - 4^2 + 7^2$
   7 A room is 4 metres square. A carpet 3 metres square is laid on the floor.
      What is the area of the part not carpeted?
   8 What fraction of a complete revolution is: 90°   45°   30°   40°   60°   180°?

58

Graphs

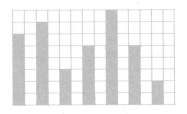 

A We have drawn box-pictures and line graphs to show information quickly and easily, as in the graphs below.

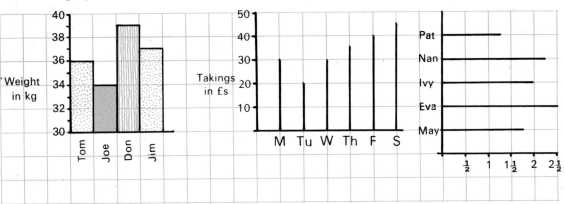

Boys' weight   Shop takings for a week   Distances (metres) in the long jump

1 Use these graphs to answer the questions below:

   (1) Who was the lightest boy?
   (2) How much heavier was Don?
   (3) What do you think was early closing day at the shop?
   (4) What were the takings on the busiest day?
   (5) Who won the long jump?
   (6) How far was the second longest jump?
   (7) How much further did May jump than Pat?

2 Now draw graphs on squared paper to show these facts:

   (1) Marks (out of 50): Tom 20, Jean 35, Bill 15, Jim 40, Betty 45, May 25.
   (2) Long jump (metres): Alan 3, Tom 2·5, Joe 3·5, Jim 3, Dick 4.
   (3) The numbers in your class with red, brown, black or fair hair.
   (4) The number of periods you spend on each subject at school.
   (5) The heights and weights of boys or girls in your class.
   (6) (a) The number of boys who would like to become—a policeman, a pilot and so on.
       (b) The number of girls who would like to become—a nurse, a secretary and so on.
   (7) The pocket money of each of these boys and girls: Allan 30p, Mary 25p, Peter 20p, Jim 35p, Ian 40p, Nan 45p.

B   Here is another kind of line graph.

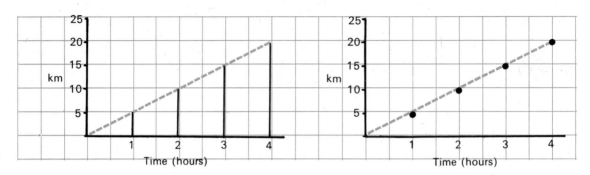

A man walks at an average speed of 5 km per hour, so that he walks 5 km in 1 hour, 10 km in 2 hours, 15 km in 3 hours and so on. You can see that the tops of each vertical column are in a straight line. So we would use a "dot" to show where the top should be and draw a straight line to join all the dots.

1 From the graph find out how far he went in $\frac{1}{2}$ hour, $1\frac{1}{2}$ hours, $3\frac{1}{2}$ hours.

2 How long would it take him to go 7·5 km, 17·5 km?

3 Draw 2 graphs like these to show a boy cycling at an average speed of 20 km an hour. Use it to find how far he would go in $3\frac{1}{2}$ hours, and how long it would take him to cycle 50 km.

4 In the two graphs shown, we were actually making up a 5 times table.

    1 × 5 = 5   2 × 5 = 10   3 × 5 = 15 and so on.
and 5 ÷ 5 = 1   10 ÷ 5 = 2   15 ÷ 5 = 3 and so on.

Now draw graphs to show some of the other tables.

C 1 Here is another kind of graph, a Pictograph, which shows a "picture story"— the marks for a class out of 80.

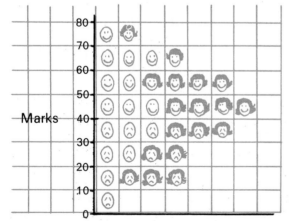

(a) How many boys in the class?
(b) How many girls?
(c) How many altogether?
(d) How many passed (smiling faces)?
(e) How many failed (sad faces)?
(f) What fraction of the girls passed?
(g) What fraction of the boys passed?
(h) What was the pass mark?
(i) Draw a box picture and a line graph (different colours for boys and girls) to show the same information.
(j) Make a pictograph about one of your class tests.

2 Can you draw a policeman?
   You can make a drawing of one like this:
   Suppose we let this drawing stand for 1 policeman.

   How many policemen would this row stand for?

Here is a Pictograph showing the number of police in county forces.

 stands for 10 policemen   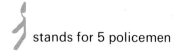 stands for 5 policemen

Redshire	🚶🚶🚶 𐌏
Blueshire	🚶🚶🚶🚶🚶 𐌏
Brownshire	🚶🚶🚶🚶🚶🚶 𐌏
Pinkshire	🚶🚶🚶🚶🚶
Greenshire	🚶🚶🚶 𐌏
Whiteshire	🚶🚶 𐌏

(a) Which county has most policemen?
(b) Which county has fewest policemen?
(c) Could you give us a reason for this difference?
(d) How many policemen are in the Blueshire force?
(e) How many more policemen has Brownshire than Greenshire?

3 Below are the numbers of bottles of milk drunk by pupils in various classes in Deanside School.

Class	Number
1	36
2	40
3	42
4	44
5	38
6	36
7	40

Let   stand for 4 bottles
and   stand for 2 bottles.

(a) Draw a pictograph to show these numbers.
(b) Find out the numbers in your own school and draw a pictograph.

4 Draw a pictograph in the proper colour to show the numbers in your class with red, brown, black or fair hair.

5 Draw a pictograph to show the number of cars, lorries, buses, vans, cyclists and walkers passing your school during 15 minutes, using pictures like these shown:

D Another way to show information simply is a **Pie Chart**.
Here is a pie cut into slices.

Here is a circle, like the top of the pie, cut into 8 "slices", showing the numbers in Class 6 (40 pupils) with red, brown, black and fair hair.

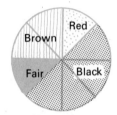

(a) What fraction of the class has red hair?
(b) How many have red hair?
(c) Which is the most common colour of hair?
(d) How many have black hair?
(e) Is it true that the number with brown hair is twice the number with black hair?

E Below are two more "pie charts", each divided into equal slices to make it easier to work out what each slice stands for.

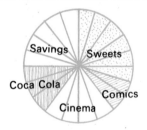

1 Jean's Pocket Money (60p)

(a) Into how many parts is the circle divided?
(b) What fraction did she spend on sweets?
(c) What did she spend least on?
(d) What did she spend on sweets and coca-cola?
(e) What did the cinema cost her?
(f) What fraction of her money did she save?

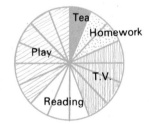

2 Jim's evening (6 pm–10 pm)

(a) What part of an hour does each small division show?
(b) How long did he play?
(c) Was this twice as long as he read?
(d) How long did he take for tea?
(e) How long did he watch T.V.?

3 Here is the number of hours May spent one day on various things.
    Sleep 10   School 6   Play 6   Eat and Wash 2

Draw a pie chart, divided into 24 equal slices to show this.

4 Find out how many periods a week you spend on Maths, English, History and Geography, Art and Music, and draw a pie-chart to show this.

5 Draw pie-charts to show how you spend   (a) your pocket money   (b) your day.

Co-ordinates

A

Mary	Jean	Bill	Tom	Ray
Joe	Betty	Hugh	Nora	Pat
Alec	Dick	Pam	Kate	Dora
Phil	Bob	Dan	Rita	Lena

Door   Teacher

Here is the plan of a class-room.

Teacher gave each of the children a code number to show where they sat.

Jean was (2, 4) which meant she was in the second column from the door and in the fourth row — or "2 along, and 4 up".

1 In the same way, can you give the code numbers for:
   (a) Nora   (b) Kate   (c) Dick   (d) Dan   (e) Joe   (f) Phil?

2 Who were given the following code numbers:
   (a) (3, 2)   (b) (1, 4)   (c) (5, 3)   (d) (3, 3)   (e) (4, 4)?

B In the same way on squared paper, we can fix the position of various points, starting from 0 at the bottom left: (where we always start counting).

So

A is (3, 2) — "3 along and 2 up"
B is (5, 3) — "5 along and 3 up"
C is (5, 0) — "5 along and none up"
D is (0, 4) — "none along and 4 up"

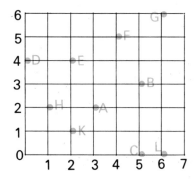

1 Now write down the code-numbers (or **Co-ordinates**) of the other points.

2 What do you think is the code number for the point 0?

C On squared paper mark off a grid like the above from 0 to 6 along and from 0 to 6 up. On it plot (mark) these points and join them in order with straight lines from A to B, from B to C, and so on. Draw a straight line from H to A.

Point	A	B	C	D	E	F	G	H
Code	(3,0)	(4,2)	(6,3)	(4,4)	(3,6)	(2,4)	(0,3)	(2,2)

What sort of shape have you made?

63

Now draw a shape of your own on a grid like this (or bigger) and by calling out the code-numbers to your neighbour, see if he can copy your shape.

D On this grid are the halves of four shapes.

Mark off a grid like this on squared paper, and then draw the half shapes as shown.

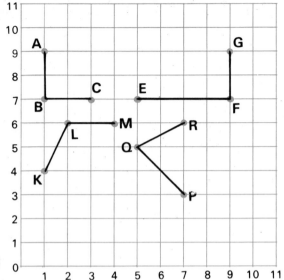

1 Mark a point D and then draw lines to make the shape ABCD a square. What are the co-ordinates of point D?

2 Mark a point H, to make shape EFGH a rectangle. What are the co-ordinates of point H and also of the points E, F and G?

3 Mark a point N, to make shape KLMN a parallelogram. What are the co-ordinates of point N and also of the points K, L and M?

4 Mark a point S, to make PQRS a kite shape. What are the co-ordinates of point S?

E Mark off another grid.

1 If Q is the point (3, 6), R is the point (5, 6), T is the point (1, 1) and S is the point (3, 1), what is the shape QRST?

2 If A is the point (7, 7), B is the point (5, 2), and C is the point (9, 2), what shape is the figure ABC?

F Mark off another grid up to 12 along and 10 up.

Now plot these points:

A	(1, 1)		I	(8, 5)
B	(2, 3)		J	(10, 8)
C	(2, 4)		K	(8, 3)
D	(0, 4)		L	(8, 1)
E	(2, 6)		M	(6, 1)
F	(2, 7)		N	(7, 3)
G	(4, 6)		P	(3, 3)
H	(4, 5)		Q	(3, 1)

Draw lines to join each point in order... A to B, B to C, and so on, and join Q to A at the end.

Put a dot at the point (2, 5). What does your shape look like?

More quadrilaterals

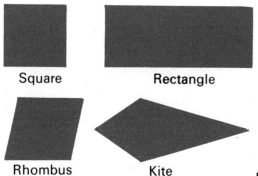

Square
Rectangle
Rhombus
Kite

We saw that a quadrilateral was a shape which had 4 sides.

We looked at the square, the rectangle, the rhombus and the kite.

Write down as many facts about these 4 shapes as you can.

Not let's look at some more members of the **quadrilateral** family.

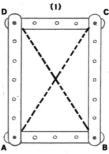 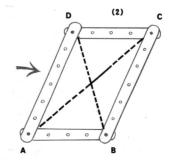

A   The Parallelogram

1 If you make a rectangle from strips of Meccano, or wood, or peg-board, and push it to the right as shown, you can form a **parallelogram**.

The dotted lines show diagonals (which can be made from elastic).

(a) Which sides are equal in each figure?
(b) Are the diagonals equal in the rectangle?
(c) Are the diagonals equal in the parallelogram?
(d) What can you say about the "corner" angles in figure (1)?
(e) What happens to the "corner" angles in figure (2)?
(f) In figure (1) the diagonals divide each other into 2 equal halves. Check this by measuring.
Do the diagonals in figure (2) "bisect" each other in the same way?

2 Now make some parallelograms in these three ways:

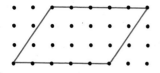  

Nailboard or pegs and peg board with elastic.

Draw on squared paper.

Fold a paper rectangle once. Then fold again. Open out and cut along dotted lines.

MATHS 4–E

3 On a cardboard rectangle make a mark (A) on the top edge.
   Now mark a point B on the bottom edge, so that B is the same distance
   from the bottom right corner as A is from the top left.

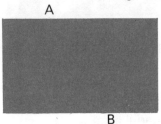 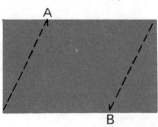 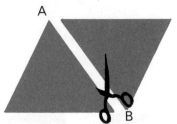

Now draw lines from A and from B to the corners as shown and cut along
these lines to make a parallelogram.
Draw a line from A to B and cut along this line to divide the parallelogram
into two triangles.
Can you now fit one of these triangles exactly over the other?
Are these triangles congruent?
What does this tell you about the sides and angles of a parallelogram?

4 Make another cardboard parallelogram in the same way, but this time
   draw in both diagonals and cut along them to make 4 triangles.
   Write on them 1, 2, 3 and 4.
   Which triangles are congruent? (Test by fitting over each other).
   Now re-arrange these triangles to form the parallelogram.
   Make a drawing on squared paper of a parallelogram and shade in one colour
   two congruent triangles, then shade in another colour the other two
   congruent triangles.

5 Draw these on squared paper, following the pattern of the squares.

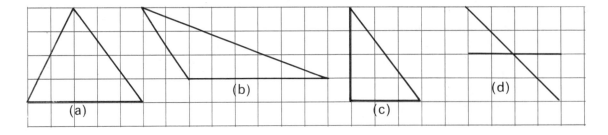

(a) Add 2 lines to each triangle to make a parallelogram.

(b) Could you add these 2 lines in more than one way?

(c) How many ways can you find?

(d) Add 4 lines to figure (d) to make a parallelogram.
     What does this show you about the diagonals of a parallelogram?

B **The Trapezium**

1 Look at the coloured shape in each of these pictures.

 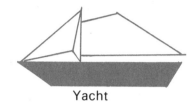 

House   Yacht   Flower pot   Hat

Are they quadrilaterals?
Are any sides parallel in each shape?

Trapezium

Isosceles trapezium

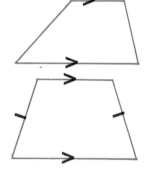

Quadrilaterals like these which have only two opposite sides parallel are called trapeziums.
If the two sides which are not parallel are equal the shape is called an **isosceles trapezium**.

(a) Has the isosceles trapezium any lines of symmetry?
(b) Which angles are equal in the isosceles trapezium?

2 Here are 3 triangles.
By drawing a line parallel to the bottom side we can make a trapezium. Which figure shows an isosceles trapezium?

Equilateral   Right-angled   Scalene

3 Fold a paper rectangle down the middle. Fold again.
Open out and draw lines as shown.
Cut along them to make a trapezium.

4 Take two squares and two rectangles of the same breadth.

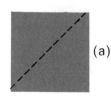   

(a)   (b)

Cut the square along a diagonal as shown. Form a trapezium with the two triangles and the rectangle.

Cut the rectangle along a diagonal and form a trapezium with the two triangles and the square.

C The Arrow-head

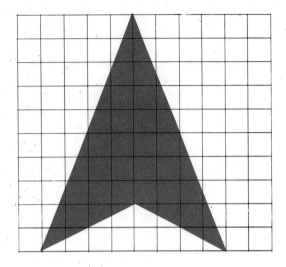

1 Here is a drawing of the arrow-head shape.

Copy it on squared paper.

Could you draw a line to divide it into two congruent triangles?

Would this be a line of symmetry?

2 Fold a rectangle of thin cardboard longways down the middle.

Cut as shown by the dotted lines.

Open it out to show an arrow-head.

Clearly the fold is a line of symmetry, as you can fold the two triangles to fit exactly.

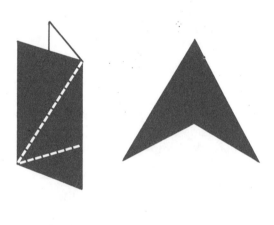

D The Scalene Quadrilateral

The last of the family of quadrilaterals is the scalene quadrilateral, which has none of its sides equal or parallel.

(a) Are any of its angles equal? Test by measuring.

(b) Are its diagonals equal?

Symmetry

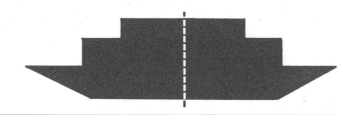

In Book 3 we looked at lines of symmetry which divided a shape into 2 equal congruent parts. Here is the letter H with 2 lines of symmetry shown.

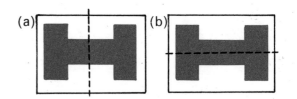

If you folded each of these about the dotted line, would each of the two halves fit exactly over each other?

Draw the letter H as above on 2 cards, one having a vertical line of symmetry and the other a horizontal line. Cut along the dotted lines.

Hold each part in turn close to a small mirror, like this:

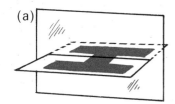

You should see that the H is completed by the image in the mirror.

Do the same with other letters.

1 Now draw the mirror image of these half-shapes.
Copy the half-shapes on squared paper and put in the dotted "mirror line".

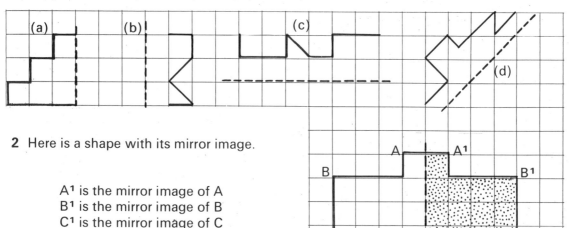

2 Here is a shape with its mirror image.

A^1 is the mirror image of A
B^1 is the mirror image of B
C^1 is the mirror image of C

(a) Notice that on the previous page A is 1 square to the left of the line of symmetry and A¹ is 1 square to the right.
What can you say about B and B¹, and C and C¹?

(b) Notice, too, that the line B¹C¹ is the mirror image of BC.

(c) Notice, too, that the line joining A and A¹ is divided into 2 equal parts by the line of symmetry.

(d) We can show this in another way. Copy the figure on squared paper.
Fold the paper about the line of symmetry, so that the shaded half goes behind.
Now use a pin and prick through the points A, B and C. Open out.
You should find the pin has pricked a hole in A¹, B¹ and C¹.

3 Here are some half-shapes. Copy them into your exercise book.

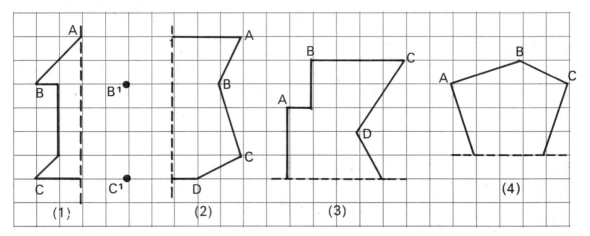

Without drawing the mirror image of each figure, can you put a dot at the point where you think the mirror image of each of the points A, B, C and D will be? (the first one is done for you).

Now answer these questions:

(a) In fig (2) A is 3 boxes to the right of the line of symmetry
      B is 2 boxes ,,   ,,   ,,   ,, ,,   ,,   ,,   ,,
      C is 3 boxes ,,   ,,   ,,   ,, ,,   ,,   ,,   ,,
      D is 1 box   ,,   ,,   ,,   ,, ,,   ,,   ,,   ,,

What then, will be the position of the images A¹, B¹, C¹, D¹?

(b) In fig (3) A is 3 boxes above, B is 5 boxes above, C is 5 boxes above and D is 2 boxes above the line of symmetry.
Where will the mirror images of each point be?

(c) What can you say about the mirror images of A, B, C, in fig 4?

Sets

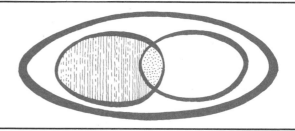

A A set can mean "things we are talking about". Suppose we are talking about the kinds of fish. Then clearly we are not talking about horses or sweets, or birds. Here is a set of farmyard animals in the farmyard.

How many sets of farmyard animals are there?
Each of these sets is a member of the set of farmyard animals, but there are several sets within this set. We can show this by drawing a large ring to stand for the whole set of farmyard animals and small rings inside it for each set of animals.

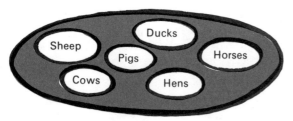

Each set—sheep, cows, hens, and so on—belong to the same bigger sets, the set of farmyard animals.

B A tea set consists of cups, saucers, plates, and so on. So in the tea set we could have a set of cups, a set of saucers, a set of plates etc, like this.

When we divide (or sub-divide) a larger set into smaller sets, each of the smaller sets is said to be a **SUBSET** of the larger set. The set of cups, the set of saucers, the set of plates, are **subsets** of the tea-set.

The set of cows, the set of horses and so on are **subsets** of the set of farmyard animals.

71

Here are some larger sets and some sets which could belong to the larger sets (that is, they are **subsets** of the larger set).

Can you pick out the larger sets to which the subsets belong?

    Cats   **Spoons**   **Dolls**   **Sparrows**   **Hammers**   **Maths**
    School subjects   **Birds**   **Animals**   **Tools**   **Toys**   **Cutlery**

1 Look at this set of numbers $A = \{1, 2, 3, 4, 5.\}$
This set has a lot of subsets. Here are two.

    $\{1, 3, 5\}$ = the set of odd numbers less than 6
    $\{2, 4\}$ = the set of even numbers up to 4.

Can you find any more subsets?

2 See if you can write down some subsets of these larger sets.

    (a) { wild animals }   (e) { dogs }
    (b) { British coins }   (f) { numbers that divide evenly by 3 }
    (c) { toys }   (g) { fractions }
    (d) { footballers }   (h) { four-sided figures }

3 Here are some subsets of larger sets. Can you write down the name of a larger set to which these might belong?

    (a) { tomatoes }   (e) { squares }
    (b) { tulips }   (f) { chocolates }
    (c) { Tom, Dick, Harry }   (g) { 2, 4, 6 }
    (d) { girls }   (h) $\{\frac{1}{2}, \frac{1}{4}, \frac{1}{8}\}$

C Here is a set of toys $T = \{\text{dolls, teddy-bears, tops, trains}\}$

and **D** is a set of Dolls $D = \{\text{dolls}\}$

You can see that the set of Dolls (**D**) belongs to the larger set of Toys.

    That is Dolls is a subset of Toys.

We can use a short way of saying this.

D ⊂ T   **⊂** means "is a sub-set of"

D Here are some sets:

B = {Britons}   G = {Girls}   D = {Daffodils}   T = {toffees}
F = {flowers}   S = {Sweets}   M = {monkeys}   P = {Scots}
E = {2, 4, 6, 8}   J = {Joan, Jean}   O = {1, 3, 5, 7}   A = {animals}

Which of these sentences are true, and which are false?

(a) D ⊂ F   (e) B ⊂ P   (i) 8 ⊂ E   (m) Rose ⊂ F
(b) M ⊂ A   (f) John ⊂ J   (j) Betty ⊂ G   (n) Caramels ⊂ T
(c) D ⊂ S   (g) 5 ⊂ O   (k) Horse ⊂ A   (o) Cheese ⊂ S
(d) 4 ⊂ E   (h) T ⊂ S   (l) 7 ⊂ E   (p) 10 ⊂ E

E Here are some mathematical sets:

A = {5, 10, 15, 20, 25, 30} = the set of numbers up to 30 that will divide evenly by 5.

B = {21, 28, 35, 42, 49} = the set of numbers between 20 and 50 that will divide evenly by 7.

C = {10, 12, 14, 16, 18, 20} = the set of even numbers from 10 to 20.

Write down in your exercise books some subsets of these 3 sets—
for example, even numbers, odd numbers, numbers divisible by 3, by 4, by 6, by 7.
Give each subset a code letter and write your answer down, using the symbol ⊂.

F Here is a set of shapes.

Pick out 3 subsets in this set, give them a code name and write down a subset sentence about each.

G 1 Here are more ring pictures showing a larger set and a subset.

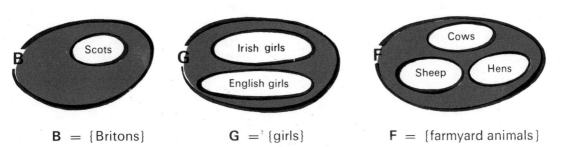

B = {Britons}   G = {girls}   F = {farmyard animals}

This shows in a picture that:

{Scots} ⊂ {Britons}

{Irish girls, English girls} ⊂ {Girls} and

{sheep}, {cows}, {hens} are subsets of the set of farmyard animals.

2 Can you say, in each of these pictures, what the coloured part stands for?
(Think of other members which may belong to the same set).

H Now draw some ring pictures to show these facts:

(a) {Daisies} is a sub-set of {flowers}
(b) {Lions} ⊂ {wild animals}
(c) {Saucers} ⊂ {tea-dishes}
(d) {4} is a sub-set of {even numbers}
(e) {Swallows} ⊂ {birds}
(f) {Pennies} ⊂ {British coins}

In each case colour in the rest of the ring picture, and write down what this coloured bit could mean.

I Here are 2 more ring pictures.

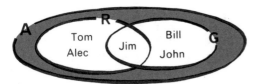

A = {Boys in my class}
R = {Boys with red hair}
G = {Boys who wear glasses}

B = {girls in my class}
P = {girls who play the piano}
Q = {girls in the school choir}

1 Now answer these questions:

(a) Which boys have red hair?
(b) Which boys wear glasses?
(c) Does Tom wear glasses?
(d) Is John red haired?
(e) What can you say about Jim?
(f) What does the coloured part mean?

2 Now answer these questions:

(a) Which girls play the piano?
(b) Which girls are in the school choir?
(c) Can Jane play the piano?
(d) What can you say about Peg?
(e) What does the coloured part mean?

J Now look at these sentences and write down if they are true or false.
Remember ∈ means "is a member of" ∉ means "is not a member of".
(1) Trout ∈ {fish}
(2) Bus ∉ {boats}
(3) 2 ∈ {even numbers}
(4) 3 ∉ {fractions}
(5) Peter ∈ {girls}
(6) Glasgow ∉ {French cities}

Volume

Everyone knows that a cupboard can hold only so much and that a jug, a bottle or other container, will take only so much water or milk and then overflow. The exact amount of room in each is called the VOLUME.

Here is a brick and a box into which the brick fits exactly. So they take up almost the same amount of space.

The brick is a *solid* and the box is a *container*.

For solids—*volume* means the amount of space they take up.
For containers—*volume* means the amount of sand, water, and so on, that they can hold (or contain) when full.

Since solids and containers are of different sizes, we need some standards for measuring volume.

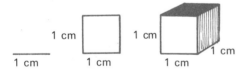

The line is 1 cm *long*
The square has an *area* of 1 square centimetre (1 cm^2)
The cube has a volume of 1 cubic centimetre (1 cm^3)

The centimetre cube is so small it is not easy to make. It is much easier to make a cube each of whose sides is 10 cm long.

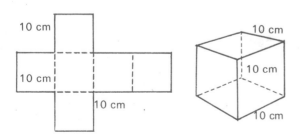

A On a piece of thin cardboard draw the figure on the left. Cut it out and fold along the dotted lines. Fasten the figure together with Sellotape to make a cube.
The length, the breadth, and the height of the cube are each 10 cm

Here are 2 cubes made up of several 1 cm cubes.

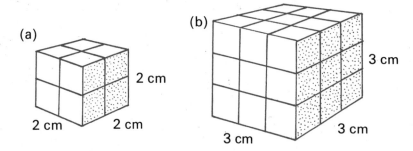

We can build up each of these cubes in "layers" of 1 cm cubes, like this:

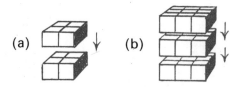

You can see that there are two layers of four 1 cm cubes in (a) and three layers of nine 1 cm cubes in (b).

We say that the volume of cube (a) is 8 cubic cm, and that the volume of cube (b) is 27 cubic cm.

Or, in other words, volume of cube (a) = $(2 \times 2 \times 2)$ cm^3 = 8 cm^3
volume of cube (b) = $(3 \times 3 \times 3)$ cm^3 = 27 cm^3

cm^3 is short for "cubic centimetres"

A cube of sides 2 cm is called a 2 cm cube.
A cube of sides 3 cm is called a 3 cm cube.

1 What is the value of $2^3, 3^3, 4^3, 5^3, 10^3$?

2 What is the volume of a 4 cm cube, a 5 cm cube, a 10 cm cube?

3 If you had some hollow cube boxes, or containers, which you filled with water or sand, what volume would fill these containers:
    (a) a 2 cm cube   (b) a 4 cm cube   (c) a 5 cm cube   (d) a 10 cm cube?

4 Make a 10 cm cube. Put your cube together with those of your friends to make shapes. What are the volumes of the shapes you made?

No doubt you found in 3 (d) above that a 10 cm cubic container has a volume of 1 000 cm^3. (It could hold one thousand of the small centimetre cubes.)

Instead of saying it has a volume of 1 000 cm³, we can say it has a capacity of 1 LITRE.

1 LITRE = 1 000 cubic centimetres

A cube like this would hold 1 000 cubic centimetres of water or 1 *litre* of milk.
A litre bottle of water, or lemonade, or milk, when full will fully fill a 10 cm cube container.

5 Write as fractions of a litre: (a) 500 cm³ (b) 250 cm³ (c) 750 cm³ (d) 125 cm³

For smaller measures the litre is divided into 1 000 parts called millilitres (for short ml)

$$1 \text{ litre} = 1\,000 \text{ ml} \qquad 1 \text{ ml} = \frac{1}{1\,000} \text{ litre}$$

A chemist uses this small (ml) measure for liquid medicines, and your mother may add so many ml of water or milk when she is baking.

Weight of 1 litre of water
Weigh an empty litre bottle or other litre measure.
Pour into it 1 litre of water and weigh again.
From this find the weight of 1 litre of water.
Did you find it weighed 1 kilogramme? What does 1 ml weigh?

6 Some boxes or cartons are in the shape of a **CUBOID**, like this:

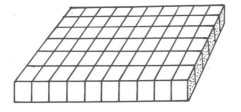

This figure shows a number of wooden 1 centimetre cubes neatly arranged in the form of a **CUBOID**.

You will see that:

(1) there are 10 cubes in each row.
(2) there are 6 rows.
(3) there are 10 × 6 = 60 cubes altogether.
(4) the volume of the cuboid is 60 cubic centimetres (60 cm³).

7 With some cm cubes make a rectangular block 5 cm long, 4 cm wide and 3 cm high.

(a) How many cubes are there in each layer of your block?
(b) How many layers are there?
(c) How many cm cubes are there altogether in your block?
(d) What is the volume of the block?

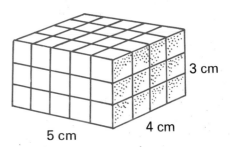

8 Make another rectangular block 7 cm long, 6 cm wide and 4 cm high.

(a) How many cubes did you need?   (b) What is the volume of the block in cubic cm?

Here is a box 6 cm long, 5 cm wide and 3 cm deep. It is to be filled with 1 cm cubes.

(a) How many 1 cm cubes are there already in the box?
(b) If another layer of cubes were placed on top of the bottom layer, would the box be full?
(c) How many layers of cubes could be put into the box?
(d) How many 1 cm cubes could the box hold?
(e) What is the volume of the box?

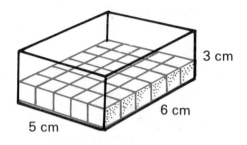

9 A box is 10 cm long and 6 cm wide.

(a) What is the area of the bottom?
(b) How many 1 cm cubes would be required to fill the bottom?
(c) If the box is 4 cm high, how many layers could be placed inside the box?
(d) What is the volume of the box?

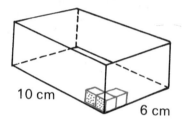

10 The table below shows the measurements of some boxes. Copy the table and fill in the missing measurements.

	Length	Breadth	Height	Volume
(a)	8 cm	3 cm	2 cm	—
(b)	12 cm	4 cm	3 cm	—
(c)	4 m	3 m	$2\frac{1}{2}$ m	—
(d)	4 cm	—	2 cm	24 cm³
(e)	—	3 cm	4 cm	60 cm³
(f)	6 m	4 m	—	120 m³

78

Solids

Above you see a man cutting logs. He is cutting right across them.

A "cut across" is called a "cross-section". You **can** see that the cross-section of the logs is a circle.

If you cut straight across any cylinder, the cross-section is a circle.

Use a round stick of plasticine as a log and cut across it.

So far we have looked at the outside of solids. Let's now have a look inside solids, in this way.

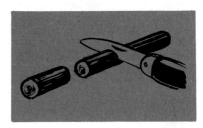

A Prisms

1 Here is a cube which is being cut across. What is the shape of the cross-section?

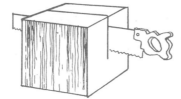

cross-section

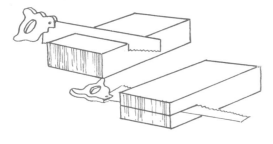

2 Here a cuboid is being "sliced" in two ways.
What shape is the cross-section in each case?
Some packets of margarine are this shape.
Name some other objects which have a rectangular cross-section.

3 Here is another solid.

(a) Do you remember what it is called?
(b) What is its cross-section if it is cut as shown?

4 You can see, so far, that if we cut right across a prism, the cross-section is the same all the way through.
We have already seen that prisms could be built up by layers of congruent shapes—squares, circles, triangles and so on.

What shapes can be made with cross-sections, or layers of:

(a) Congruent squares? (b) Congruent rectangles?
(c) Congruent triangles? (d) Congruent circles?

5 Many pencils have an end which has 6 sides.

(a) What name do you give to a shape with 6 sides?
(b) What shape would the cross-section be each time, if cut as shown?
(c) Is this solid a prism?
(d) How do you know if it is a prism?

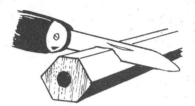

6 In your exercise book, draw the outline of the cross-sections of these objects, if they were cut straight across:

(a) a rubber   (b) a milk-straw   (c) a plank
(d) a banana   (e) a Toblerone packet (f) a piece of chalk.

Which of these are not prisms? Why?

7 Write down the names of some objects which have these cross-sections:

   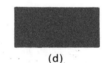 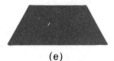

(a)   (b)   (c)   (d)   (e)

8 Arrange 9 equal cubes in 3 rows of 3 to fit together. Imagine a slice taken off the top cutting across from left to right.

(a) Make a drawing on square paper of what the top of the slice would look like.

(b) Make a sketch of what the cross-section would look like if
   (1) the middle cube was removed   (2) a corner cube was removed.

9 In shops you can see sliced loaves, sliced meat and so on which show cross-sections. Can you name any more like these?

B **Pyramids**

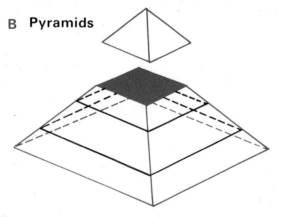

1 Here is a **Square Pyramid**.
(a pyramid with a square base)

If you cut across it, as shown, you can see that the cross-section is a square, but the square gets smaller and smaller as you get nearer the top.

Shapes like these squares are said to be similar (but not congruent).

The square pyramid can be built up from layers of similar squares, which get smaller and smaller from bottom to top until the squares get so small that the last one "disappears".

So we could almost call a pyramid a "Disappearing Solid".

2 If the bottom of the pyramid was a rectangle, what would be the cross-section if we sliced across?
Would the cross-sections be similar or congruent?

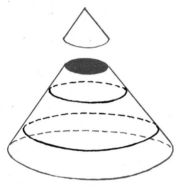

3 Here is a cone.
  (a) What shape is the cross-section, if sliced as shown?
  (b) In what way is the cone like a pyramid?
  (c) Could you call a cone a "Disappearing Solid".

C Spheres

Spheres, as we saw, are solids like footballs, tennis-balls and so on. Many oranges, peaches and sweets are almost spherical.
If you cut straight across a sphere in any direction, what shape is the cross-section?
Cut across an orange.
What shape is the cross-section?

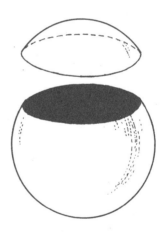

D Solid from Solid

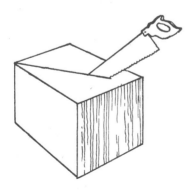

Here is a cuboid. If it is cut down, as shown, what 2 solids are now formed?
When you cut the cuboid like this, you will get 2 triangular prisms which can be fitted together to make a larger triangular prism.
Try this with a cuboid made of plasticine if you can.

MATHS 4-F

E An Odd Solid

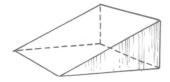

1 Here is a solid called a **Wedge**.

(a) What can a small wedge be used for?
(b) What food is often cut in this shape?
(c) What kind of solid would you say it is?

2

 (a)

 (b)

If you slice it this way, the cross-section is a right-angled triangle.
So the wedge can be built up of layers of congruent right-angled triangles.

So it is a PRISM!

If you slice it this way, the cross-sections are rectangles of different sizes which are similar. So it can be built up of layers of similar (but not congruent) rectangles.

But since it can be built up from layers of congruent right-angled triangles, we call it a TRIANGULAR PRISM.
Can you think of any other "odd" solids like this?

3 Here are 4 cones "sliced" through at different angles.

(a)   (b)   (c)   (d)

(1) What is the shape of the cross-section in each case?
(2) What is the name of the curve forming the outer edge of the cross-section in cone (d)?
(3) In which cone do you slice off a smaller one?
(4) How many surfaces has each part of the cones after slicing?
(5) How many of these surfaces in each part are curved and how many are flat?

Making sure 3

A 1 Find: a ($\frac{1}{4}$ of £72) + ($\frac{1}{3}$ of £54)   b ($\frac{1}{3}$ of £9·36) − ($\frac{1}{10}$ of £15·60)
       c ($\frac{1}{8}$ of £13·20) + ($\frac{1}{2}$ of £1·06).

   2 Write the following, putting in the correct sign (=, > or <) in place of the question mark:
      (a) (12 × 5) ? (7 × 8)
      (b) (10 × 10) ? (25 × 4)
      (c) (18 × 6) ? (16 × 7)
      (d) (200 ÷ 4) ? (350 ÷ 7)
      (e) (300 ÷ 6) ? (387 ÷ 9)
      (f) (25 × 5) ? (960 ÷ 10)

   3 A motorist went on a journey of 252 km and travelled at an average speed of 56 km an hour. If he set off at 8.55 am, at what time did he arrive?

   4 Find the distance travelled in $4\frac{1}{2}$ hours at an average speed of:
      (a) 48 km/h   (b) 56 km/h   (c) 64 km/h   (d) 72 km/h   (e) 84 km/h.

   5 Write the following as decimal fractions:
         50%   25%   80%   65%   75%   5%

   6 Find: (a) 25% of £60   (b) 75% of 32 francs   (c) 30% of $50   (d) 5% of 80¢

B 1 A boy had 20 sweets. He has already eaten 6 of them:
      (a) What decimal fraction of the sweets has he eaten?
      (b) What percentage has he eaten?
      (c) What percentage has he left?

   2 Find the value of **n** in these equations:
      (a) 3n + 7 = 25   (b) 5n − 3 = 22   (c) $\frac{2}{3}$ of n = 6   (d) $\frac{5}{8}$ of n = 15

   3 Find: (a) $3^2 + 4^2$ (b) $6^2 \div 3^2$ (c) $8^2 \div 4^2$ (d) $5^2 \times 3^2$ (e) $9^2 - 7^2$

   4 Find the volume of these cuboids:

	Length	Breadth	Height		Length	Breadth	Height
(a)	6 cm	4 cm	2·5 cm	(b)	4 m	3 m	1·5 m

   5 Here is the plan of a room with a carpet.
      (a) What is the area of the room?
      (b) What is the area of the carpet?
      (c) What is the area of the floor not carpeted?
      (d) What is the perimeter of the room?

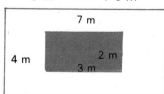

C 1 Find: (a) 0·75 of £12 (b) 0·45 of £20 (c) 0·25 of 30 m (d) 0·85 of 60 km.

2 A car uses 9 litres on a journey of 108 km. How far does it go on:
   (a) 1 litre   (b) 3 litres   (c) 20 litres   (d) 35 litres?

3 Change these decimal fractions to percentages.
   0·5   0·6   0·25   0·75   0·45   0·4   0·88

4 A road map is drawn to the scale of 1 cm to 10 km. How many km do these lines stand for:
   (a) 1·5 cm   (b) 2·4 cm   (c) 6·8 cm?

5 Here are the distances from Edinburgh to six other towns. Using the scale of 1 cm to 10 km, draw lines to stand for these distances:

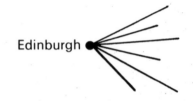

   Glasgow 70 km   Lanark 53 km   Biggar 46 km
   Dunbar 45 km   Moffat 83 km   Peebles 37 km

6 Five men weigh 64·360 kg, 65·650 kg, 66·550 kg, 67·100 kg, and 63·580 kg. What is their average weight?

D 1 Give approximate answers to the following:
   (a) 798 × 5   (b) 889 ÷ 3   (c) £3·92 × 4   (d) 98 cm × 6   (e) $8·95 ÷ 3

2 If a town had an average of 10 hours' sunshine daily for the first 3 days in August, how many hours' sunshine did it have in the whole month (31 days)?

3 A, B and C are three towns. B is 35 km due north of A, and C is 55 km due east of A. Take a scale of 1 cm to 10 km and make a scale drawing like the sketch. Now find the distance from town B to town C.

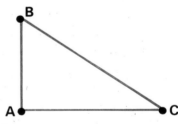

4   a = 4   b = 6   c = 10   d = 12
   Use these numbers to find the value of the following:
   (1) a + b + c   (2) a + c − b   (3) b + c − a   (4) c + d − b
   (5) $\dfrac{a + b}{c}$ (6) $\dfrac{b + c}{a}$ (7) $\dfrac{a + b + c}{d}$ (8) $\dfrac{b + c + d}{a}$

5 How many degrees does the minute hand of a clock pass through in going:
   (a) once round the clock   (b) from 8 am to 8.15 am   (c) from 8.15 to 8.45 am
   (d) from 8.15 to 9 am   (e) from 8 to 8.10 am   (f) from 8 to 8.20 am
   (g) from 8.10 to 8.15 am   (h) from 8.30 to 8.55 am   (i) from 8 to 9.10 am?

Work cards

1 Here are some prices taken from the MENU of a restaurant in New York.

Chicken Pie	$1.65	Sausage & Eggs	$1.25
Ham Steak	$2.25	Spaghetti	95¢
Roast Beef	$1.50	Apple Pie	25¢
Turkey	$1.85	Sundaes	55¢
Pork Cutlets	$1.65	Ice-cream	15¢
Liver & Bacon	$1.95	Coffee	20¢

How much would a person pay for the following, and how much change would he receive from a $5 bill?

(a) Ham Steak, apple pie, coffee
(b) Turkey, ice-cream, coffee
(c) Sausage and eggs, coffee
(d) Chicken pie, sundae, coffee
(e) Roast beef, ice-cream, coffee
(f) Liver & bacon, sundae, coffee
(g) Pork cutlet, ice-cream, coffee
(h) Spaghetti, apple pie, ice-cream

2 Arrange the numbers up to 50 in rows like this

1	2	3	4	5	6	7	8	9	10
11	12	13	14	15	16	17	18	19	20

Draw (a) a small square round each square number,
(b) a small rectangle round each rectangular number,
(c) a small triangle round each triangular number.

1. Which numbers are both square and rectangular numbers?
2. Which numbers are both triangular and rectangular?
3. Are there any numbers which are triangular and square numbers?
4. Which numbers are neither square nor rectangular? What are these numbers called?

3   (a) In a cinema the rows of seats in the stalls are lettered A to W.
       There are 22 seats in each row. How many seats are there in the stalls?

    (b) The rows of seats in the balcony are lettered A to L. In each row there are 20 seats.
        How many balcony seats are there?

    (c) One night there were 29 empty seats in the stalls and 17 in the balcony.
        How many people were in the cinema?

    (d) 10% of the people were children. How many adults were there?

4   A ladder 5 metres long was placed against a wall. The foot of the ladder was 3 metres from the wall. How far up the wall did the ladder reach?

    To find out make a scale drawing.
    Choose a scale of 1 cm to 1 metre.

    (a) Draw 2 straight lines at right angles to each other.

    (b) Make AB 3 cm long.

    (c) Open your compasses 5 cm.
        With centre B draw an arc to cut the vertical line at C.

    (d) Join BC.

    (e) Measure AC in cm.
        AC = **** cm. So ladder reaches **** m up the wall.

    Now find in the same way how far these ladders reach up the wall:

	Length of ladder	Distance from foot of wall
(1)	6·5 m	2·5 m
(2)	7·5 m	4·5 m

5   Measure carefully each line in this drawing and make a copy of the drawing exactly twice the size.

    If your drawing is a scale drawing of a field and each cm stands for 40 m, what is the distance all round the field?

    Find also the area of the field.

6 On a 10 cm cardboard square draw this figure.
Letter it as shown and cut out the 7 shapes.
Use these shapes to make:

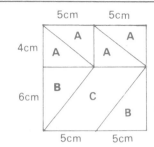

1. a rectangle from 2 of shape A.
2. an isosceles triangle from 2 of shape A.
3. an isosceles triangle from 2 of shape B.
4. a parallelogram from 2 of shape A.
5. a parallelogram from 2 of shape B.
6. a parallelogram from 2 of shape B and shape C.
7. a rhombus from 4 of shape A.
8. a trapezium from shape C and 1 shape B.
9. an isosceles trapezium from shape C and 2 of shape B.
10. a kite from 2 of shape A and 2 of shape B.

7

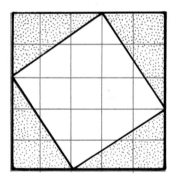

1. Each small square stands for 1 square cm.
   (a) What is the area of the large square?
   (b) Are the 4 triangles congruent?
   (c) What is the area of each triangle?
   (d) What is the area of the uncoloured square?

2. In this box there are 6 layers of eggs. There are 12 eggs in each row and there are 6 rows.

   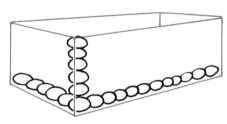

   (1) How many eggs are there in the box?
   (2) How many smaller boxes would be required if they were packed:
   (a) 6 to a box?   (b) in dozens?   (c) 18 to a box?
   (3) If 75% of the eggs are large eggs, how many small eggs are there?
   (4) If a box of this size, full of eggs, is sold at 1 new penny per egg, what would be the value of the whole box?

8

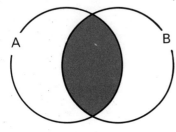

(1) (a) Write down all the odd numbers less than 12.

(b) Write down all the numbers up to 12 which divide evenly by 3.

(c) Use a **10p coin to draw two intersecting circles.** Name them A and B.

(d) In circle A put in all the odd numbers you have written down.
In circle B put in all the numbers divisible by 3 which you have written down.
Don't write any number more than once.

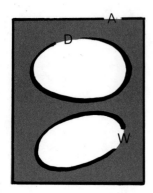

(2) Draw a rectangle and in it draw two rings. Name them A, D and W.

A = {all 4-legged animals}

D = {dogs}

W = {wild animals}

Now put the following in their proper set:

| collie | lion | tiger | spaniel |
| terrier | wolf | bulldog | elephant |

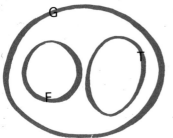

(3) Draw these three rings:

G = {all things which grow in the ground}

F = {flowers}

T = {trees}

Now put the following in their proper set:

| oak | daisy | tulip | elm |
| lily | rose | ash | beech |

9 Cut some strips of cardboard of various sizes. Pin them to a board with drawing pins to show various shapes—the triangle family, the quadrilateral family, regular polygons. Perhaps you should work in a group to do this.

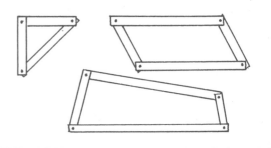

10

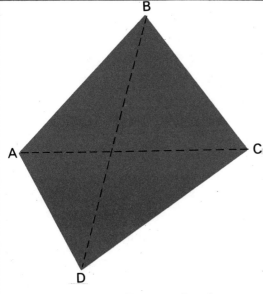

1 This is the plan of a field, drawn to a scale of 1 cm to 20 metres.

(a) What is the actual length represented by:

0·1 cm   0·5 cm   0·75 cm   0·9 cm?

(b) Measure each side on the plan and then give the approximate length of each side of the field.

(c) What is the perimeter of the field (approximately)?

(d) Find as accurately as you can the distances across the field from A to C and from B to D.

2 Here are the distances by air, to the nearest 100 km, from London to 6 different parts of the world. Using a scale of 1 cm to 1 000 km (1 mm to 100 km), draw straight lines to stand for these distances:

Aden 6 600   Bombay 7 800
Copenhagen 1 000   Montreal 5 300
Rio de Janeiro 8 800   Marseilles 1 200

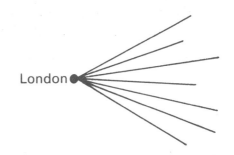

11   We can use symbols to stand for numbers.

If △ = □ □ □ □, then △ △ stands for □ □ □ □ □ □ □ □.

1 If ○ stands for □ □, how would you show □ □ □ □ □ □?

2 If ★ stands for □ □ □, what would ★★ stand for?

3 What other way could you show ★★ ?

4 What would these equal? △ + ○, △ − ○, $\frac{△}{○}$, $\frac{△△}{○○}$

5 We can make up secret messages using symbols.
If these symbols stand for soldiers, and 🚶🚶🚶🚶🚶 = ★
what would this mean? "Send ★★★★★ at once".

Now make up some symbol messages of your own.

12

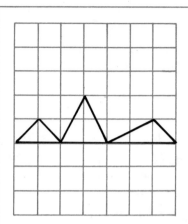 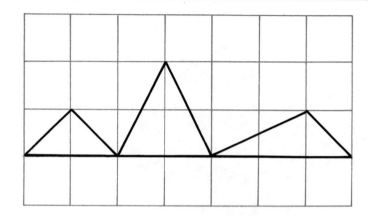

Here we see a way of making a drawing bigger (or **magnifying** it) by using squared paper.

**Each line in the larger drawing is twice the length of each line in the smaller drawing.
So the scale is twice the scale of the smaller drawing.**

Figures like these which are the very same shape but different in size are called **similar figures**.
Are they congruent?

1. Now draw the figures below and then draw similar but larger figures. Make them twice the size by using 2 boxes for every box in the smaller figures.

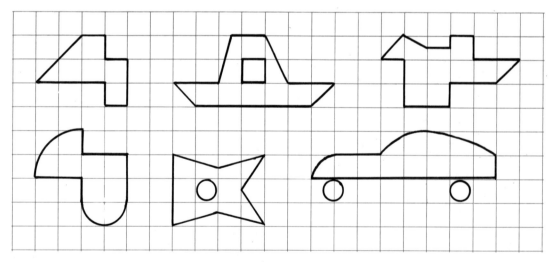

2. Make some larger drawings on squared paper and then draw similar ones half the size.

13

   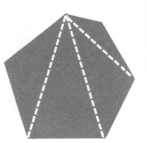

Above are some shapes divided into triangles.

Draw shapes like these of 3, 4, 5, 6, 7 and 8 sides. The sides don't need to be of the same size. Divide them into triangles always starting at the same point. Count the number of triangles in each shape and enter your results in a table like this:

Number of sides	3	4	5	6	7	8
Number of triangles	1					
Number of diagonals	0					

What do you notice about the results?

Could you now say into how many triangles you could divide shapes with:
(a) 9 sides?   (b) 10 sides?   (c) 12 sides?

14

X	1	2	3	4	5	6	7	8	9		
1	1	2	3	4							
2	2	4	6								
3	3	6									
4	4	8									
5											
6											

Make up a multiplication square up to 12 times 12.

Put a ring round the number 30 wherever you see it in the table.

You should find, using row and column, that:

$30 = 10 \times 3 = 3 \times 10 = 5 \times 6 = 6 \times 5$
3, 5, 6 and 10 are called **factors** of 30.

Find in this way some factors of:
12   18   21   24   36   42   64

15

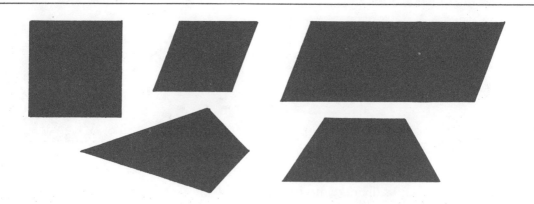

Copy this table. Tick off what facts are true about each shape above.

	■	◢	▰	◆	⬢
All sides are equal	✓				
The opposite sides only are equal					
Only 1 pair of angles is equal					
Has only 1 line of symmetry					
The diagonals are equal	✓				
The diagonals cross at 90°					
The diagonals divide the shape into 4 congruent triangles					

16 This table gives the length, breadth and height of 6 cuboids.
Copy the table and fill in the volume:

	Length	Breadth	Height	Volume		Length	Breadth	Height	Volume
1	8 cm	3 cm	2 cm		4	4 m	3 m	0·5 m	
2	6 cm	$2\frac{1}{2}$ cm	2 cm		5	2·5 m	2 m	2 m	
3	5 cm	4 cm	$2\frac{1}{4}$ cm		6	3 m	2 m	1·5 m	

17 Lattice Multiplication

This was used in Italy about 500 years ago.
Let's multiply 234 by 567.
Put 234 across the top, and 567 down the side.
Draw diagonals as shown.
Fill in the "lattice" by multiplying $2 \times 5, 3 \times 5, 4 \times 5$ in the top row, and so on.
Now start with the bottom right hand corner, and put down 8.
Add diagonally, "carrying", if necessary, to the next diagonal on the left, as shown.
The answer is 132 678.

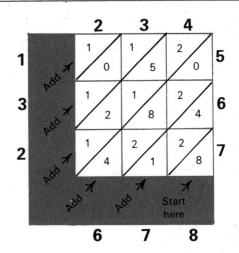

Note: 1×5 is written as

1 Use this lattice way to multiply:

(a) 27 × 35   (b) 48 × 69   (c) 87 × 92
(d) 243 × 562 (e) 337 × 428 (f) 813 × 964

2 Make up some multiplication sums of your own and use the lattice way.

18

  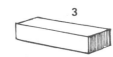

I am a girl. You will find my name if you write correctly, and in order the following letters of the names of the solids shown above.

Christian name	Letter		Solid
a	4th		5
b	2nd		2
c	Last	of	5
d	Last		1
e	Last		4
f	2nd		1

Surname	Letter		Solid
a	3rd		1
b	2nd		5
c	3rd	of	4
d	3rd		2
e	2nd		3
f	Last		1
g	5th		1

Make up names of your own and see if your neighbour can find them.

19 **How Squares and Cubes grow**

1 Use squared paper and colour in a square.
Let each side of this square stand for 1 cm.
Now draw a thick line to show a square with side 2 cm (length = 2 boxes).
Continue in this way up to a square of side 10 cm.

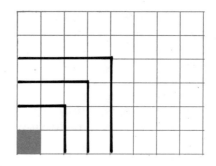

Make a table like this up to a square of side 10 cm.

Side (cm)	Area (cm²)	Index Form
1	1	1^2
2	4	2^2
⋮	⋮	⋮
10	100	10^2

2 Make a table like this to show how cubes grow.
Continue to side of 10 cm.

Side (cm)	Area of one face (cm²)	Volume of Cube (cm³)	
1	1	1	1^3
2	4	8	2^3
3	9	27	3^3

20 Copy these diagrams. Write in the size of all the other angles.

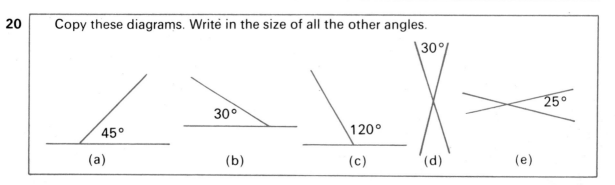

The Tangram

One of the oldest puzzles in the world is the Chinese Tangram. It was very popular with the Chinese long before Christ was born.

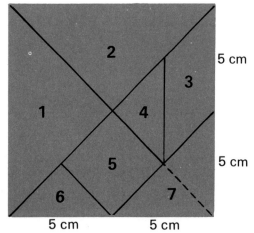

Take a 10 cm cardboard square and mark it off as shown. Measure carefully. Number the parts as in the drawing. Cut out the square and then cut along the lines you have drawn. Don't cut along the dotted line. You will now have 5 triangles, a parallelogram and a square.

Close this book and then try to make the following shapes:

1. Use two triangles to form a square.
   Now change the square into a parallelogram.
   Have you changed the area?

2. Use three pieces to form a rectangle.
   Now change the rectangle into a parallelogram.
   Have you changed the area?
   How do you find the area of a rectangle?
   Can you now think of a rule for finding the area of a parallelogram?

3. Make a trapezium with 3 pieces.

4. Form a parallelogram with 4 pieces.

5. Use the square, the parallelogram and two small triangles to form a trapezium.

6. Use 3 pieces to form a triangle.

7. Put the 7 pieces together to form the original square.

8. Put the 7 pieces together to form a large rectangle.

9. Put 5 smaller pieces together to form a large triangle.

For centuries people have amused themselves making shapes with the Tangram pieces. Copy the shapes below and then make others for yourself.

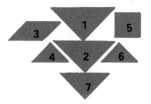 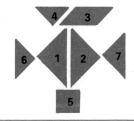 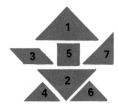

Useful information

Until we go completely metric you will sometimes want to know how metric weights and measures compare with the weights and measures commonly used at present. Here are the approximate equivalents:

Length:
- $2\frac{1}{2}$ centimetres — 1 inch
- 30 centimetres — 1 foot
- 1 metre — 3 inches more than a yard
- 1 kilometre — $\frac{5}{8}$ of a mile.

Weight:
- 1 kilogramme — a little more than 2 lb.
- 500 grammes — about 1 lb.

Capacity:
- 1 litre — $1\frac{3}{4}$ pints
- $4\frac{1}{2}$ litres — 1 gallon
- 5 millilitres — about a teaspoonful
- 10 millilitres — about a tablespoonful

Money:

New Money	50p	10p	5p	2p	1p	$\frac{1}{2}$p
Old Money	10/–	2/–	1/–	5d	2d	1d

Here is a quick way of changing shillings and pence to new pence and vice versa. (The answer is not always exact, but is useful as a quick guide.

Example 1
Change 16/6 to new pence
Leave out the stroke—166
Divide 166 by 2
So 16/6 = 166 ÷ 2 = 83p

Example 2
Change 12p to s d
Leave out the p and multiply by 2
12 × 2 = 24
Put in the stroke—2/4
So 12p = 2/4